DESIGNING SMART OBJECTS IN EVERYDAY LIFE

DESIGNING SMART OBJECTS IN EVERYDAY LIFE

Intelligences, Agencies, Ecologies

Edited by
Marco C. Rozendaal, Betti Marenko and William Odom

BLOOMSBURY VISUAL ARTS
LONDON • NEW YORK • OXFORD • NEW DELHI • SYDNEY

BLOOMSBURY VISUAL ARTS
Bloomsbury Publishing Plc
50 Bedford Square, London, WC1B 3DP, UK
1385 Broadway, New York, NY 10018, USA
29 Earlsfort Terrace, Dublin 2, Ireland

BLOOMSBURY, BLOOMSBURY VISUAL ARTS and the Diana logo are trademarks of Bloomsbury Publishing Plc

First published in Great Britain 2021

Cover design: Louise Dugdale

A catalogue record for this book is available from the British Library.

Library of Congress Cataloging-in-Publication Data
Names: Rozendaal, Marco C., editor. | Marenko, Betti, editor. |
Odom, William (Designer), editor.
Title: Designing smart objects in everyday life: intelligences, agencies, ecologies / editors, Marco C. Rozendaal, Betti Marenko, William Odom.
Description: London; New York: Bloomsbury Visual Arts, 2021.
Identifiers: LCCN 2021004545 (print) | LCCN 2021004546 (ebook) |
ISBN 9781350160125 (HB) | ISBN 9781350160132 (eBook) | ISBN 9781350160149 (ePDF)
Subjects: LCSH: Internet of things. | Cooperating objects (Computer systems) Classification: LCC TK5105.8857 .D488 2021 (print) |
LCC TK5105.8857 (ebook) | DDC 006.2/2–dc23
LC record available at https://lccn.loc.gov/2021004545
LC ebook record available at https://lccn.loc.gov/2021004546

ISBN: HB: 978-1-3501-6012-5
ePDF: 978-1-3501-6014-9
eBook: 978-1-3501-6013-2

Typeset by Newgen KnowledgeWorks Pvt. Ltd., Chennai, India
Printed and bound in Great Britain

CONTENTS

ILLUSTRATIONS

Figures

Table

LIST OF CONTRIBUTORS

Philip van Allen is a Professor at ArtCenter College of Design in the Media Design Practices MFA program. He is developing new models for AI interaction, including non-anthropomorphic 'animistic intelligence'. He also develops tools for prototyping AI and works with industry and government through his Commotion.ai consultancy. He has been a recording engineer, technologist, entrepreneur and researcher, working across entertainment, industry, start-ups and education. He received his BA in experimental psychology/cognitive science from the University of California, Santa Cruz.

Kristina Andersen is an Assistant Professor of Future Everyday/Industrial Design at the Technical University of Eindhoven, TU/e. She has a long history of making with, alongside and through the notion of, magic machines. Her work is increasingly concerned with how we can allow each other to imagine things through digital craftsmanship and collaborations with semi-intelligent machines.

Evert van Beek is a PhD researcher and designer at Delft University of Technology. He has a background in industrial design engineering with a focus on interaction design (MSc from Delft University of Technology) and experience in the field of human-computer interaction (HCI) (Mintlab KU Leuven). His doctoral research involves product and service innovation in the energy transition addressing issues related to participation, experimentation, active things and engaging with data.

Arne Berger was born in the seventies in the Eastern Bloc. He is an interaction design researcher and left handed. His background is in media arts and design, with an interdisciplinary master from Bauhaus-University Weimar. Officially a computer scientist with a doctorate in engineering from a technical university, his research takes an inter- and transdisciplinary research-through-design approach at the intersection of co-design and interaction design. He explores novel modes of co-creation together with people in the context of their homes and neighbourhoods. Arne is a Professor of human-computer interaction at Hochschule Anhalt (the second home of the Bauhaus).

Maaike Bleeker is Professor of Theatre Studies at Utrecht University. In her work she combines approaches from the arts and performance with insights from philosophy, media theory, STS and cognitive science. She is also an experienced dramaturg. Bleeker is research leader of "Acting like a Robot: Theatre as Testbed for the Robot Revolution" (2021–5) and the author of (among others) *Visuality in the Theatre: The Locus of Looking* (Palgrave 2008) and *Doing Dramaturgy: Thinking Through Practice* (Palgrave 2021). She (co-) edited several volumes including *Transmission in Motion: The Technologizing of Dance* (Routledge 2016).

Nazli Cila is an Assistant Professor in human-agent partnerships at the Delft University of Technology. Her work focuses on investigating the collaborations humans make with autonomous agents, such as robots, smart products or AI systems, and the socio-technical implications of these collaborations. She holds a PhD in human-centred design from TU Delft and is interested in integrating empirical work (i.e. experimentation, future modelling, and prototyping) with the philosophical, ethical and practical issues regarding trust, responsibility, control and intelligence.

Jelle van Dijk is Assistant Professor in Human-Centred Design at the University of Twente, the Netherlands. He trained as a cognitive scientist and did his Phd in industrial design at Eindhoven University of Technology. He investigates the embodiment and situatedness of human beings in relation to the design of interactive technology. In this regard he also researches how participatory and do-it-yourself design methods may help people to get a better grip on their own lifeworlds, focusing in particular on people that challenge majority norms.

Carl DiSalvo is an Associate Professor at the Georgia Institute of Technology, with appointments in the School of Interactive Computing and the School of Literature, Media, and Communication, and he directs the Experimental Civics Studio. His work combines methods and theories from design, the social sciences and the humanities to explore the social and political qualities of computing. His first book, *Adversarial Design*, is part of the Design Thinking, Design Theory series at MIT Press. He is also a co-editor of the MIT Press journal *Design Issues*.

Christopher Frauenberger is Professor for HCI at the Center for Human-Computer Interaction, Paris-Lodron University of Salzburg, Austria. He received his PhD in computer science from Queen Mary, University of London, and holds a habilitation (*venia docendi*) in informatics from TU Wien, Austria. His research interest lies in understanding and designing for the entangled relationship between humans and digital technologies, with a particular concern for their political, ethical and ecological implications. He is inspired by philosophy and committed to a participatory design practice.

Pim Haselager is an Associate Professor at the Department of Artificial Intelligence and a principal investigator at the Donders Institute for Brain, Cognition and Behaviour, both at the Radboud University in Nijmegen. He focuses on the ethical and societal implications of cognitive neuroscience and AI. He publishes on the societal implications of AI and cognitive neuroscience in journals such as *Nature: Biotechnology*, *Science and Engineering Ethics*, *American Journal of Bioethics*, *Neuroethics*, *Journal of Cognitive Neuroscience* and *Journal of Social Robotics*.

David Kirk is Professor of HCI and Director of Open Lab and the Centre for Digital Citizens in the School of Computing at Newcastle University, UK. He has a background in psychology (BSc), ergonomics (MSc) and HCI (PhD) and is a chartered psychologist and Fellow of the Royal Society of Arts. Much of his human-centred design research focuses on the design of technologies for domestic spaces. He is particularly inspired by philosophical anthropology.

Ann Light is Professor of Design and Creative Technology, University of Sussex, UK, and Professor of Interaction Design, Social Change and Sustainability, Malmo University, Sweden. She has worked on democratizing futures and the politics of technology with arts and grass-roots organizations and marginalized groups on five continents, using participatory and co-design methods. Her interests include design for sharing, social justice in the digital economy and how creative practice can support transformations to sustainable living. Over many years, she has addressed digital networks and is the co-author of *Designing Connected Products: UX for the Consumer Internet of Things* (O'Reilly 2015).

Betti Marenko is a transdisciplinary theorist, academic and educator working across process philosophies, design studies and critical technologies. She is the author of numerous articles, book chapters and essays, and co-editor of *Deleuze and Design* (Edinburgh University Press 2015). She is the founder and director of the Hybrid Futures Lab, a transversal research initiative developing speculative-pragmatic interventions at the intersection of philosophy, design, technology and future-crafting. She is Reader in Design and Techno-Digital Futures at Central Saint Martins, University of the Arts London (UAL), and WRHI Specially Appointed Professor at Tokyo Institute of Technology.

Effie Le Moignan is a research associate at Open Lab, Newcastle University, UK. She completed her PhD in HCI at Northumbria University (2018), focusing on the family snapshot in an Instagram context. Her research focuses on the domestic sphere and visual culture, with a focus on how emergent and future domestic technologies can be influenced and understood by their analogue and historical predecessors. She is interested in how humanness can be manifested, expressed and provoked by the use of everyday technology.

William Odom is an Assistant Professor in the School of Interactive Arts and Technology at Simon Fraser University in Vancouver, Canada, where he co-directs the Everyday Design Studio. He leads a range of projects in slow interaction design, the growing digitization of people's possessions and methods for developing the practice of research-through-design. He holds a PhD in HCI from Carnegie Mellon University and was previously a Fulbright Scholar in Australia, a Banting Fellow in Canada and a Design United Research Fellow in the Netherlands.

Tobias Revell is a digital artist and designer from London, programme director at the London College of Communication, UAL, and co-founder of design research consultancy Strange Telemetry, critical technology outfit Supra Systems Studios and approximately 47.6 per cent of research and curatorial project Haunted Machines. He lectures and exhibits internationally on design, technology, imagination and speculation. He is a PhD candidate in design at Goldsmiths.

Dries De Roeck has a background in industrial design and has been switching between the academic research and design practitioner hats ever since. He is a designer and researcher, with a strong interest in how technology impacts the day-to-day life of people. He holds a joint PhD degree in social sciences (KU Leuven) and product development (UAntwerpen) where he is also wrapping up his PhD research. Dries is a board member of ThingsCon, a leading community of IoT practitioners in Europe. He co-organizes the family friendly hacker camp Fri3d Camp and organizes technology-related activities for primary schools.

Marco Rozendaal is Associate Professor of Interaction Design at TU Delft's Faculty of Industrial Design Engineering in the Netherlands, where he directs the Expressive Intelligence Lab. With a background in interactive media, design and engineering, his research straddles multiple disciplines and combines practical, critical and methodological perspectives. Marco's current work explores the design of new interaction styles and paradigms engendered by artificial intelligence. In his work, he is strongly committed to bringing design research to a broader audience through exhibitions and events.

David Verweij is a creative technologist and a doctoral trainee at Newcastle University, UK. Originally taught as an interactive product and experience designer at Eindhoven University of Technology (BSc, MSc), he develops novel and bespoke technologies for research. His doctoral research addresses the digital engagement disparity for smart home products amongst the family.

PREFACE

What happens when your doorbell starts talking to your thermostat behind your back? When your fridge decides to add items to your weekly shopping list? When your favorite jumper, the kitchen table and the kettle seem to manifest signs of digital personality?

We are surrounded by objects that are increasingly 'smarter'. They are embedded in connected systems and operating largely behind our conscious attention. The growing connectivity and smartness of everyday objects raises questions concerning what counts as 'intelligence' in a context in which objects seem to possess some digital forms of it. Besides, established notions of objecthood are being superseded by the changing nature of digital objects. The question 'what is an object?' becomes more complicated to answer when we consider smart objects whose potential to act, function and respond to data-rich environmental inputs is far greater when compared to their more 'passive' analogue counterparts. The potential range of adaptations exhibited by smart objects can make them act in ways that go beyond goals of task-oriented efficiency to enact behaviours that may exceed, and perhaps even confound, people's expectations. In short, smart objects are on a trajectory to become our everyday companions, collaborators, co-inhabitants, and even co-conspirators.

This collection of essays, descriptions of empirical work, and design cases brings together perspectives from interaction design, the humanities and science and technology studies (STS) to map, explore and interrogate ways in which people's relations to everyday smart objects can be expanded and reimagined. Broadly, the aim of this volume is to propose a shift from a human-machine interaction framework to the notion of increasingly multistable ecologies between and among human and non-human actors. This book's overall goal is to explore and critically reflect on the possibilities and consequences bound to the growing landscape of smart objects that increasingly populate our everyday lives.

This book is largely the result of an interdisciplinary workshop that was held at the Lorentz Center, University of Leiden, the Netherlands, from 30 April to 4 May 2018. As researchers, academics, practitioners and educators at the intersection of design, HCI, computing, STS, and the humanities, the organizers Marco

Rozendaal, Betti Marenko, William Odom and Kenny Chow's intention was to foster the development of a research agenda to support the interdisciplinary field of interaction design in investigating the 'smartness' of everyday objects from a multiplicity of perspectives. The overall aim of the workshop, and now of the present collection, is to offer a practical-theoretical scaffolding to shape and extend interaction design practice and research. Our hope is that this book can provide the interaction design community with a diversity of frameworks, approaches and strategies to guide the design of everyday smart objects in valuable, critical and responsible ways.

INTRODUCTION

The increased integration of computation and networking capabilities into physical products is transforming many of our everyday objects into smart ones. Things such as domestic appliances, furniture, clothing and toys are gaining new capabilities and expanding their modes of interaction with their users. This prompts a series of questions concerning their role and agency: the way in which they may be perceived by the users and how their extended capabilities shape and inform the way they are designed. How are smart everyday objects ontologically different from their analogue counterparts? How are their new identities shaped by people's perceptions, experiences and imaginations? More crucially for the scope of our inquiry, how do we design them? What are the new frameworks, strategies and practices that can inform the design of smart everyday objects?

One of the recurrent themes anyone investigating smart objects has to contend with is the amount of debate (and hype) that has surrounded the Internet of Things (IoT) since its beginning. The expression 'Internet of Things' has been in circulation for two decades; the networked systems it describes were studied and experimented upon since the early 1980s (Sterling, 2005; Greenfield, 2006; Greengard, 2015). However, it can be said that the full potential of the interconnected world is yet to be reached. It is telling that the 'things' in IoT have been framed and described in varied ways including smart entities (Kuniavsky, 2010) and enchanted objects (Rose, 2014), or as non-human actors expressing a variety of possible personalities (Marenko & van Allen, 2016) and existences (Wakkary *et al.*, 2017).

For instance, when smart objects are perceived to have an autonomous existence, they might be experienced as well behaved, acting on our behalf, or as bossy, arrogant, mischievous or even incompetent when things don't go as expected. This variety of personalities can challenge the design of such objects. People are used to many contemporary everyday products responding to our actions, such as a kitchen mixer where we push a button or turn on a switch and the machine whirrs into action, blending or performing some other function. Yet, this scenario is considerably different once we consider smart kitchen appliances in the context of IoT (Atzori, Iera & Morabio, 2010) and ambient intelligent environments (Aarts & Wichert, 2009). What happens when our appliances talk back, take initiative and perhaps even advise us on what to eat? When they become more thoughtful, adaptive, social, suggestive and even capable to question our choices? How is their identity – and potential social roles – shaped, interpreted and designed for, considering ethical issues about responsibility, accountability and agency?

It is clear that when working with smart objects we have to consider their identity and character (Laurel, 1997; Janlert & Stolterman, 1997; Govers, Hekkert, & Schoormans, 2003), their qualities across a gradient of the human and non-human (Levillain & Zibetti, 2017), and the crossing of the virtual and physical they embody (Sterling, 2005). Thus, the proposition we put forward here is that smart objects in everyday life are a blend of tools and agents, a hybrid of the human and the non-human, possessing emergent properties and different forms of agency, and therefore demanding a different definition of 'intelligence'. Smart objects challenge interaction designers to grasp and creatively work with the new opportunities for design they can offer (Giaccardi, Speed, Cila & Caldwell, 2016), to use data or artificial intelligence (AI) as new materials in interaction design practice (Holmquist, 2017; Dove, Halskov, Forlizzi & Zimmerman, 2017; Odom & Duel, 2018) and to reimagine interaction as collaboration, coexistence and cohabitation with humans (Marenko, 2014; Marenko & van Allen, 2016; Rozendaal, 2016).

A further challenge for interaction designers is to envision smart objects from the perspective of their networking capabilities, therefore in terms of the wider ecologies in which they are embedded and in which they function, rather than from the perspective of a single entity (Funk, Eggen & Hsu, 2018). For instance, take smart kitchen appliances aware of their environment and of each other, and therefore able to act in concert. You can imagine a situation in which the blender knows that a third egg is not needed in a specific recipe you are using to make a cake. While you are unaware of this and attempt to take that additional egg from the fridge, the fridge (having received information from the blender) might decide to lock its door to prevent you from taking that egg. What may happen next? Perhaps, it is now the blender that issues a counter-order. One can begin to imagine how operational conflicts might emerge among machines that are, by design, enabled to be 'opinionated'. Likewise, one may wonder how far these networked ecologies may reach. Should the blender and the fridge be talking with your car too, now planning a trip to the supermarket? How do we deal with the distributed nature of an ecology of smart objects as designers as well as end users? To which extent do, could or should these ecologies reach? What might be the implications for the emergence and inherent unpredictability of responses in extended ecologies populated by multiple, active and potentially diverging agents?

The kitchen appliance example above shows how it is important, necessary and urgent to propose a shift from the conventional user-object relationship to wider ecologies of the human and the non-human where the actors engaged (whether they are people, objects or data) affect each other in negotiable, situated and intelligent ways. This landscape requires new design frameworks, perspectives, approaches and methods to help us (as designers, as well as users) consider, critically reflect on and rethink how smart objects are experienced and designed. Smart objects require innovative, hybrid and transversal methodologies to be contextually understood in their form, appearance and behaviour (Hoffman, Kubat & Breazeal, 2008; Hoffman & Ju, 2014; Vallgårda, 2014; Rozendaal, Ghajargar, Pasman & Wiberg, 2018), to be experimented upon and prototyped in everyday life (Chamberlain, Crabtree, Rodden, Jones & Rogers, 2012; Odom *et al.*, 2012; Desjardins, Viny, Key & Johnston, 2019), and, crucially, to be imagined and speculated about in their future possible manifestations (Auger, 2014; Wakkary *et al.*, 2015; Oogjes, Odom & Fung, 2018).

This is what the present book is about: a collection of insights, reflections and propositions to build a research agenda that, drawing on a multiplicity of perspectives, can shape, extend and evaluate interaction design practice and research for the current and near future landscape of smart everyday objects. This research agenda investigates and proposes alternative perspectives on intelligence, which, by underpinning new and thoughtful ways to design smart objects and the interactions we have with them, can open up design opportunities that can leverage the growing capacities of smart objects. A core goal of this research agenda is to propose new approaches to design that enable future smart objects to be imagined, to be given form and to be prototyped and situated in people's everyday lives. Finally, this research agenda puts forward careful consideration of the impact of smart objects on our individual, collective and social everyday lives, and is informed by critical reflection on their emergent agencies and their cultural, ethical, legal and political implications (Stolterman & Croon Fors, 2008; Redström & Wiltse, 2018).

Designing smart objects in everyday life

The notion of 'Agents' is a key lens to understand and frame objects as 'smart'. Following computer scientists Michael Wooldridge and Nicholas R. Jennings's definition of agents (1995), these are entities that are autonomous, acting without a direct intervention of humans; reactive, perceiving the environment and reacting to changes in a timely fashion; proactive, being able to exhibit goal-directed behaviour by taking the initiative; and social, having the ability to interact with other agents including humans. This lens allows us to describe and explore smart objects as technical infrastructures and to analyse their physical embodiment, software and networking capability.

To start with, the physical embodiment of smart objects allows them to be responsive. Embedded sensors, such as cameras, microphones and touch sensors allow smart objects to see, hear and feel the environment and allow us to interact with them through gestures, voice or by directly manipulating the physical object. Embedded displays, LEDs, physical controllers and speakers allow these objects to communicate back to us in a visual, haptic or aural manner. The way in which these objects communicate becomes more sophisticated when they can allow physical movement or change shape. This can be made possible by embedded mechatronics or smart materials. The rapid developments in engineering and material science will enable smart objects to display such novel forms of expression to be commonplace in the near future. Processors embedded in such physical artefacts enable them to collect, store and process data captured locally as well as elsewhere, and provide them with a 'brain'.

Software 'makes' objects 'smart' by allowing them to make sense of the environment (as picked up by its sensors) and to react in ways that can produce 'intelligent' behaviour. This might be very simple, for instance, when connecting the input from a sensor to the output of an actuator. Cyberneticist Valentino Braitenberg demonstrated that by connecting the light intensity (captured by a light sensor attached to a vehicle) to the speed of the motor driving that vehicle, rudimentary forms of intelligent behaviour can be achieved (Braitenberg, 1986). AI introduces more sophistication by enabling smart objects to learn by process, to be updated, to harvest data and to develop more advanced and refined models of the world as the system continues to learn. Different types of AI or machine learning (ML) exist that vary by their type of learning and knowledge representation (Michalski, Carbonell & Mitchell, 2013). For example, AI can learn through simple instruction, through evolutionary ways and by probabilistic inferences, and the knowledge of AI's may include rules of behaviour, problem-solving heuristics or classification taxonomies.

The networking capabilities of smart objects make them extend outside their immediate environment via connectivity protocols through which objects are connected to a network of other objects and systems (Kortuem, Kawsar, Sundramoorthy & Fitton 2010; Al-Fuqaha, Guizani, Mohammadi, Aledhari & Ayyash, 2015). Embedded processors and wireless communication devices such as WiFi combined with networking protocols, such as IP, enable smart objects to project distinct identities and to move beyond the confines of their physicalities. This can provide smart objects with connected services 'in the cloud'. For example, a smart car can tap into the live data about traffic generated by other objects and systems to gain updates about the preferred direction of travel. This can also enable smart objects to function in a collective; multiple objects can be part of a meaningful human activity, as we have illustrated with the smart kitchen example.

Concerning their smartness, this networking capability also allows the intelligence of the object to be outsourced to another location or distributed across multiple objects as a form of a 'collective intelligence' (Mulgan, 2018).

If this illuminates smart objects as technical infrastructures that provide them with new capabilities, how do we understand them as being meaningful in everyday life? If we take the everyday as the sum of our everyday practices and day-to-day routines through which we humans accrue identity, relate to our environment and develop sense-making practices (Shove, 2007; D'Adderio, 2011), how might smart objects shape and transform it? Our starting point is that objects – whether analogue, digital or a mix of them – are essential to everyday practises, which are social and cultural, and they are profoundly enmeshed with objects (Kaptelinin & Nardi, 2006: Suchman, 2007; Kuutti & Bannon, 2014; Shove *et al.*, 2017) and therefore affect innumerable aspects of our working, social and intimate lives (Appadurai, 1986; Turkle, 2007). Deeply considering the everyday also requires framing it in the long-term and recognizing its repetitive cycles of use over longer periods of time which are stable and predictable but can also accommodate messiness, change and transformation (Kuijer, Jong & Eijk, 2013; Engeström, 1999).

To illustrate these notions with our previous example, kitchen appliances – whether smart or not – operate within the context of specific domestic environments, for instance, the kitchen, which possesses a situated cultural specificity conveyed by a variety of registers, from the spatial layout to the kind, size and number of various tools and appliances, as well as prevailing routines of use established by culture. Thus, any experience of, and interaction with, kitchen appliances can only be analysed in the interrelation with this environment and the other objects within it, as well as with the multiple human actors intersecting it. In other words, kitchen appliances are socially and culturally embedded. For example, as they are often available to many users, kitchen appliances tend to be socially coordinated. Furthermore, as they also embody traditions, histories and legacies that are both cultural and personal, kitchen appliances are inherently entangled with meaning-making practices and shared ways of knowing.

If our experience of the everyday is defined by socially embedded, meaning-making and culturally situated objects through which we develop long-term routines of engagement, then the experience of, and interaction with, smart objects should be wisely calibrated to fulfil these requirements – for instance, by striving to move beyond instantaneous gratification in user scenarios and embracing instead longer-term cycles of use and non-use. Succinctly, the presence of smart objects within wider environments ought to inform new modes of coexistence and cohabitation between us and them (Odom *et al.*, 2014).

To articulate how these may emerge (and be designed for), we now turn our attention to a discussion of intelligences, agencies and ecologies – the three key themes that underpin our framing of smart objects in everyday life. The next section explores how the *intelligence* of smart objects is shaped by a mix of the actual technical capabilities of an object as well as by human attributions of, and perception of, intelligent behaviours. Further, it discusses the *agency* of smart objects as a relational property emerging from our interaction with them. Lastly, it discusses how they are embedded in wider *ecologies* of the human and the non-human. To recognize the distributed, multiple and layered nature of smart objects in everyday environments we have chosen to use the plural terms – intelligences, agencies and ecologies. This underscores the idea that every time we engage with smart objects we are interacting with pervasive intelligent systems, with a multiplicity of coexisting (and not always aligned) agents, in wider ecologies of humans and objects.

Intelligences

As technological developments in AI and ML change the landscape of the design of interactive artefacts, intelligence becomes de facto a material to design with (Holmquist, 2017). Holmquist emphasizes that designers should be aware of the different types of ML and, in this, have a critical understanding of AI and the possibilities of what it can and cannot provide. Moreover, smart materials and mechatronic capabilities allow for new expressivities of smart objects through their material properties and the object's form(s). For example, smart polymers and shape memory materials have an inherent dynamic that can provide physical expressiveness in interaction design and more subtle, delicate and nuanced forms of physical interaction. Therefore, designing smart objects requires designers to have a broad understanding of what is meant by 'intelligent' objects. AI, ML and the technical infrastructures supporting networked smart objects are all crucial to this definition. So are the nuances of how humans perceive objects to be smart and attribute to them 'sentience'.

Humans have an innate tendency to attribute some kind of intelligence or sentience to inanimate things, even when we are perfectly aware that they are inanimate. As was shown as early as 1944 by psychologists Fritz Heider and Marianne Simmel, people would attribute intent to moving geometric figures and use anthropomorphic descriptions to explain the behaviour of abstract shapes – especially when objects appear to move by themselves and movement is not perceived to be caused by external forces. Brian Scholl and Tao Gao (2013) propose that this is hardwired in our perceptual system as an innate response to specific motion cues, such as self-propulsion, synchronous movements, patterns of approach or avoidance or coordinated orientation. Media theorist Cifford Nass introduced the term 'ethopoeia' to describe the attribution of humanness to computers that do not look human and are known not to be human (Nass, Steuer, Henriksen & Dryer, 1994; Reeves & Nass, 1996). More recently, cognitive and social scientist Leila Takayama (2009) explored agentic objects in the context of human-robot interaction, where objects that seem to have agency 'are perceived and responded to in-the-moment as if they were agentic despite the likely reflective perception that they are not agentic at all' (p. 239).

An understanding of animism – the attribution of liveliness to things – may be particularly useful as a perspective to interpret contemporary forms of human-machine interaction characterized by autonomous movement, environmental awareness and a range of expected (and some unexpected) responses. Developmental psychologist Edith Ackerman (2005) describes artefacts in the context of interactive toys as having an ambiguous nature, somewhere between the animate and the inanimate: 'the object's "aliveness" facilitates identification. At the same time, its "thingness" helps us keep a secure distance' (p. 1). Design theorist Betti Marenko (2014) introduced the notion of neo-animism to account for the 'new forms of cognition—embodied, sensorial, contextual and distributed—that are produced by ambient intelligence through mapping, tagging, and data gathering' (p. 223) and broadly in the wide networked entanglements of humans and digital things. Furthermore, Marenko and Phil van Allen proposed animistic design (2016) as a speculative and imaginative tool to rethink human-machine interaction 'neither from the perspective of the user, nor from the perspective of the object but from the ongoing modulation of their less-than-predictable interaction' (p. 2).

Philosopher and cognitive scientist Daniel Dennett's notion of 'intentional stance' offers an explanation as to why people's attribution of intention to objects is a fundamental aspect of human interaction with the world (1989). For Dennett, there is no difference between living or non-living things as long as using the intentional stance is an economical means to explain and predict complex behaviour. Adopting the intentional stance implies assuming that things have beliefs and desires and that things act rationally according to these beliefs and desires. How we arrive at these attributions of intelligence depends on the underlying metaphor that we adopt. Metaphors allow people to understand and communicate the workings of a system through a mental model (Lakoff & Johnson 1980; Norman, 1993; Janlert & Stolterman, 1997). A number of metaphors have been developed to understand how people make sense of and interact with different agents. For Instance, here we look briefly at biological and non-biological metaphors.

While biological metaphors are inspired by human, animal or plant life, non-biological metaphors have their origin in the expressiveness of cultural artefacts explicitly defined as 'enacted'. Human metaphors (i.e. anthropomorphizing) are apparent in the design and use of conversational agents and social robots that interact with human speech or use expressive body language (Allen *et al.*, 2001; Breazeal, 2003; McTear, Callejas & Griol, 2010; Følstad & Brandtzæg, 2017). However, when the humanlike appearance of robots prompts attributions of 'human' capacities (for instance, to feel, sense or express), this might conflict with the actual sophistication of the robot (Gray & Wegner, 2012) and induce a perception of 'uncanniness' – when they appear too lifelike (Mori, MacDorman & Kageki, 2012). This is also why animal metaphors that afford more-than-human perceptions of intelligence are often deployed for social robots (Breazeal, 2003).

As for non-biological metaphors, in conventional product design, objects are often perceived to have a personality that stems from the stylistic aspects in their design (Janlert & Stolterman, 1997; Laurel, 1997; Govers *et al.*, 2003; Boer & Bewley, 2018) or are 'enacted'. In other words, designed objects might appear to have an identity, their own social life (Appadurai, 1986) or 'objecthood' (Candlin & Guins, 2008). Alex Taylor refers to 'Machine Intelligence' (2009) as the lifelike quality of a machine's movement, autonomous interactions in and with the world around it, 'something "seeable", but also something enacted — emerging from those particular details of a setting' (p. 8). Marco Rozendaal (2016) introduced the notion of 'Objects with Intent' to describe agents that take advantage of the meaning of everyday things as the site for their intelligence and agency. These objects are approachable and intuitive in use, since their intelligence is made meaningful as everyday things with familiar uses, anticipated contexts of use and known ways of interaction.

The design of carefully calibrated interaction dynamics – accounting for human and non-human actors, and the networked systems that bind them together – can be achieved by acknowledging the ways in which technological innovation embedded in smart objects intersects how intelligence is attributed to objects. This may be achieved by sidestepping mainstream applications of human-like and animal-like metaphors in favour of more radical perspectives, such as animism. We contend that such an approach, by accounting for the wide spectrum of the animate and inanimate with no clear-cut division among them, can greatly contribute to the design of novel expressive forms and mental models, and to the production of narratives and fictions underpinning future interactions with smart objects. Thus, a question that emerges here is: Which animism-driven strategies can enable the creation of new kinds of interactivity and embodied relations with smart objects in everyday life by combining form-giving practices in product design and character animation?

Agencies

As much as intelligences and agencies are intertwined, it is useful to examine the notion of agency (or agencies) separately, to map key insights and literature on this topic as they feed into our proposed research agenda. Agency is taken here as a relational capacity that emerges through interaction or, following philosopher Karen Barad's argument, through what she calls intra-action. For Barad, while interaction assumes separate individual agencies preceding the interaction itself, intra-action acknowledges instead the emergence of distinct agencies *in* their act of coming together (2007, p. 33). This framing of agencies is particularly useful to understand complex ecosystems where humans and digital objects coexist and 'come together' in a variety of continuously modulated and 'live' ways – some overt, some invisible and happening in the background. For instance, social media status updates competing for our attention by actively prompting us on our smartphones, smart thermostats changing our home temperature depending on which dwellers are recognized to be at home, a lighting system adapting hues and tones to better suit our moods or refrigerators that automatically order more almond milk when it is predicted to be running out. These are all forms of interaction in which smart objects fed by environmental data manifest agency that respond to people's needs and wishes, all the while also informing our own human responses and reactions.

This ceaseless mutual calibration between human and non-human agencies calls for interaction models that are equally supple and negotiable. This means that, rather than understanding objects as tools that mediate our day-to-day activities, new models would see them instead as partners, companions or allies. The shift from tools to partners is an important one, as it raises questions concerning the ontological dimension of objects. If they are now active co-creators of interaction (rather than passive slabs of matter), extended throughout a live network of other connected objects (rather than a discrete singular entity), and partners (rather than servants or mere tools), then the traditional subject-object divide becomes distributed – and with this comes significant implications for the role of the subject, or user, or human in the equation. The questions are then how do such objects mediate interaction? What type of future partnerships can be envisioned in light of issues that include privacy, control, surveillance and accountability? How do different forms of intelligence lead to different modes of agency? And which roles might smart objects begin to play in our everyday lives?

The shift from tools to partners has been addressed by human-computer interaction literature, where the changing interactions between humans and computers-as-agents have been described initially as mixed-initiative user interfaces (Hearst, Allen, Guinn & Horvitz, 1999; Horvitz, 1999) and as symbiotic and integrative (Jacucci, Spagnolli, Freeman & Gamberini, 2014; Farooq & Grudin, 2016). With their growing autonomous and negotiable activity, smart objects can now be described as partners that 'construct meaning around each other's activities, in contrast to simply taking orders. They are codependent, drawing meaning from each other's presence' (Farooq & Grudin, 2016, p. 28). Similarly, the notion of co-performance, drawn from social practice theories, is also used to denote how new modes of human-computer relation develop through situated and evolving complementarity of capabilities and actions (Kuijer & Giaccardi, 2018).

Furthermore, different levels of agency that objects can display on the basis of the complexity of their perceived behaviour is described by the notion of 'Behavioural

Objects' (Levillain & Zibetti 2017) – for instance, the level of 'animacy' denotes objects that move spontaneously and show a consistent motion and trajectory over time while the level of 'agency' denotes objects that seem to have goals and are able to deal with changing environmental constraints in a flexible manner. The level of 'mental agency' indicates objects that seem to coordinate their behaviour with others, displaying communicative actions and showing varied attitudes to other agents. Similar incremental levels of agency are identified in the behaviour of objects within IoT (Cila, Smit, Giaccardi & Kröse, 2017). Here, at the lowest level objects collect and aggregate data to visualize patterns of behaviours, as demonstrated by quantified-self technologies such as Fitbit, or domestic 'helpers' such as Google Nest as an object that learns to adapt to users' behaviour patterns. On the highest level of agency, however, objects may develop creative contributions. Describing machines that 'make' robots becomes a way to speculate on robots that might develop artificial forms of self-awareness.

From the perspective of Activity Theory – a cultural-historical view on human psychology and development (Rubinshtein, 1946; Leontiev, 1975; Vygotsky, 1978) – all objects are considered to have conditional agency, which simply means that objects produce effects because of their physical manifestation (Kaptelinin & Nardi, 2006). Some objects possess delegated agency – the agency delegated to them by someone or something. Finally, only certain entities, such as human beings or animals, have need-based agency. While objects cannot have a genuine need-based agency, they may however appear to have one. Considering objects as 'quasi-subjects' (Latour, 1993; Bødker & Andersen, 2005) or 'subject-objects' (Suchman, 2011) allows us to grasp smart objects as social and communicative beings, whose capabilities are other than ours. Susanne Bødker and Peter Anderson describe a ship's automated control system as quasi-subject to which actions can be delegated within the complex activity of ship navigation and control. Lucy Suchman (2011) talks about 'subject-objects' in her project to understand the identity of social robots from a feminist philosophical viewpoint. Drawing on Bruno Latour's notions of quasi-objects and quasi-subjects – real, collective and discursive elements underpinning human social bonds (Latour, 1993, p. 89) – these ideas suggest the growing hybridity of social actors and systems where the human encounters the technical.

Another approach to agency comes from post-phenomenology, which sees agency as the way in which objects mediate, shape and influence our experience and interaction with the world around us (Ihde, 1990; Verbeek, 2005). For post-phenomenology, what matters most is not agency per se but considering technologies as 'mediators of human experience' rather than merely functional, utilitarian or as symbolic objects. In this context, humans and technologies shape each other in a mutually constitutive way. These ongoing mediations give rise to the subjectivity and objectivity of a given situation in the world. Intriguingly, by looking at technological objects as designed artefacts (and not things that come in a 'raw' form), post-phenomenology offers an important lens for designers to work with, considering the mediating qualities of the smart objects that might be created.

To sum up this section on Agencies, and how it informs our proposed research agenda, a key issue concerns the acknowledgement of the partnerships that humans form with smart objects. For this reason, it becomes essential to discern how humans and smart objects' abilities, capacities and competencies can complement each other, as they are practised and performed in everyday life. For instance, what are the tools and know-hows needed to interpret correctly the level of agency smart objects exhibit? How can tasks be shared (and delegated) among humans and objects? Furthermore, What are the salient experiences that designers need to consider – issues of control, trust and accountability – and which range of acceptable roles may smart objects play when people start to coexist with them in everyday life? What remains to be seen, then, is the extent to which agencies (both human and non-human) might align or diverge, might recognize or misunderstand each other, might have common goals and expectations in terms of a desired state to achieve in the world, and what may happen in potentially antagonistic situations.

Ecologies

In this context we define Ecologies as the wider ecosystems where smart objects coexist and interact with humans and with other objects, actors and infrastructures – both analogue and digital. Broadly, we describe these ecologies as populated by various assemblages of humans and non-humans, and characterized as 'largely uncharted design territory, ridden with complexity, diversity, opaqueness, and intangibility' (Funk *et al.*, 2018, p. 1). Notably, the term 'ecologies' intends to emphasize the profoundly contextual and pluralistic nature of the entanglement of human and non-human, and, in the specific of smart objects, the multiplicity of technologies, materialities, users, outcomes and infrastructures at different spatial and temporal scales that shape ecologies as such.

Ecological theories acknowledge not only that we are embedded in the context we inhabit but also that our physical and cognitive abilities have evolved as a product of the environments in which we dwell. In such ecological theories, the notion of 'embodiment' is critical (Dourish, 2004). In evolutionary biology, for example, the intelligence and behavioural repertoire of a given species are said to have co-evolved within the habitats or milieus of that species (Darwin, 2004). In psychology, the ecological approach proposed by James J. Gibson (1979), understand the human perceptual system to be tightly integrated with the action system (and thus with our embodied intelligence). A similar view is expressed by Rodney Brooks (1991) in his work on robot development, where intelligence is understood as consisting of multiple layers of sensory-action feedback systems tailored to the environments in which they operate.

Social sciences' Actor–Network Theory (ANT) (Latour, 2005) is also illuminating how ecologies of smart objects can be understood. ANT is a distinctive approach to social theory and research which originated in the field of science studies. It is best known for its insistence on the agency of the non-human. It examines the complex interrelations of human and non-human actors as they interact within a largely horizontal landscape. It considers all human behaviour to arise by the agglomeration of multiple 'actants', which can be humans, things or even ideas. Unlike conventional assumptions that people make things and objects, ANT takes this idea and turns it around. What if it was objects that make people? This is the shift proposed by ANT: from the traditional distinction between humans and things to a new ecosystem of human and non-human actants. Briefly, everything that exists must be regarded as an actant: all entities, be them natural, artificial, human or non-human, objective or social are actants; thus, they exercise agency as they ceaselessly enter in associations, alliances and networks with each other. The first thing that strikes in this ontology is how utterly horizontal it appears to be. Not only are that blender, this fridge, our laptops and your smartphone very real and very likely connected to each other, but they are also engaged in alliances to assert themselves as social actors, with various degrees of agency that they exert in the world. This emphasizes how any discussion of ecologies is always also a discussion of agencies. Whether digital or analogue, animate or inanimate, these agents all participate in (and exit) complex ecologies of alliances and relations.

Now if we consider smart objects as participating in multistable ecologies of relations, we might ask what kind of relations these would be. Relations among artefacts that shape an ecology of things can be distinguished on the basis of their 'purpose' (when objects are related in terms of how they are a meaningful component of everyday activities). They might also be distinguished on the basis of their 'context of use' (when objects physically and temporally coexist in a specific setting) or even on the basis of the 'meanings' they have been given that express their significance in people's lives (Jung, Stolterman, Ryan, Thompson & Siegel, 2008). Similarly, the notion of 'product ecology' (Forlizzi, 2008) is useful to understand how systems of technology-based products are socially and culturally situated among specific communities of people. It also illuminates how products are effectively used and by whom, as it takes into account how different social roles and attitudes, each with their own temporalities and flows, will inform people engagement and patterns of use.

The notion of ecologies also casts light on the process of adaptation – whereby 'the introduction of a new artefact to an ecology can influence various aspects of users' daily behaviours as well as the use of other artefacts' (Jung *et al.*, 2008,

p. 206). In a study on the introduction of a robot vacuum cleaner in the domestic environment, Forlizzi and DiSalvo (2006) found that the robot, by enabling new ways of cleaning, also altered established cleaning practices. Conversely, the robot also required assistance from the household inhabitants who had to intervene by rearranging and moving furniture for the robot to perform its tasks, or even helping it when it got stuck. The authors observed an 'unusual dynamic between the product, the physical environment, and participant' (p. 262), suggesting that the introduction of the robot in the household triggered multiple points of adaptation. The product, in other words, becomes an 'instigator for change — how it has an effect on people, place, and other products in use, effecting dynamic change on all of the factors in the Product Ecology' (Forlizzi, 2008, p. 15).

A final theme to consider in relation to ecologies concerns the emergent, and therefore potentially unpredictable, nature of the interaction. As multiple actors interact with one another as semi-autonomous entities, fed by live data picked up from different sources within their immediate environment and ambient networks, their interactions might become increasingly difficult to predict. If margins of unpredictability can be considered as an organic outcome of the emergent behaviours of complexity adaptive systems (Mataric, 1993; Callejas & Griol, 2005), the implications of 'digital uncertainty' in ecosystems populated by humans and smart objects can have dramatic consequences or, in a more mundane context, could lead to frustration, bafflement and a disruption of expectations. To go back to our initial example, think about the scenario in which your fridge refuses to be opened, or your blender decides (against your judgement) that your cake mix is now sufficiently done. A research agenda would need to consider ways to harness and maximize the creative potential of this type of emergent uncertainty to gain insights – for instance, on how to introduce elements of surprise, curiosity, wonder and delight in the design of meaningful everyday interactions.

To conclude this section on Ecologies, the key issues for our research agenda concern an enhanced sensitivity to the contexts within which smart objects operate, the assemblages they enter into with other objects and with users, and the type and nature of the relations they form. As embedded agents in wider ecologies, smart objects have to be examined (and designed) with an understanding of how their introduction in an existing ecosystem alters the equilibrium and changes existing relations. Furthermore, they have to be considered in their capacity to both adapt to and instigate mutual adaptability from other actors. Finally, the implications of emergent behaviours, such as the 'spontaneous' interplay between multiple intelligent actors, and the unpredictable scenarios that may arise must be taken into consideration, especially in their potential to supply creative elements to the design of surprise, delight and wonder in the everyday, and how to harness it.

Towards a research agenda

The book is structured in four parts – Perspectives, Interactions, Methodologies and Critical Understandings. Taken together, these parts outline a coherent research agenda for interaction design. Said agenda has two key aims: to understand the way smartness is expressed and interacted with in the everyday, and to offer a roadmap for the conceptualization, design, prototyping and realization of smart objects that are considered as intelligent agents located in ecologies shared with humans and non-humans. A broad view of what can be considered as 'intelligence' allows this research agenda to eschew anthropocentric determinism and to embrace instead a multi-perspectivism that considers how more-than-human forms of intelligence may feed in, and inform, the effective design of smart objects, from the mental models they express to the interaction and partnerships they foster. The research agenda also aims at highlighting salient issues concerning the social, ethical and legal implications of smart objects, and how to design while offering responsible and sustained value to people in their everyday environments.

The variety of voices collected throughout this volume, each with its own distinctive perspective, epistemic culture and research methods, indicates the value of transdisciplinarity. Working across disciplines is nothing new for design, but the range of positions and concerns presented here makes a compelling argument in favour of transdisciplinarity. The multiple entanglements between human and non-human intelligences and agencies, and how they both constitute developing ecologies of multiple actors, appeal to developing transdisciplinarity knowledge that transcends the natural and the artificial, the biological and the cultural, and bridges the theoretical and the practical. Taken together, the chapters that follow offer insights, reflections, inspiration and concrete concepts to inform the generation of a research agenda to work with, and contribute to, the wide debate on how interaction design can move forward in its enterprise of designing future interactions and experiences with smart objects in everyday life.

The volume does not explicitly propose tools or toolkits ready for implementation but rather offers a range of insights that can help define, envision and inspire further design practices. Its ambition is that these insights, together with clear, useful and inspiring methodologies, can be used in the process of envisioning, giving form and prototyping smart objects, and the ever-evolving interactions we are part of in our everyday lives. The collective voices in this book further suggest that empowering people through the design of smart objects requires fostering democratization and fairness in their design and development. Finally, this volume should also be read as an accurate, albeit transient, snapshot of the state-of-the-art discussion on interaction design in the European and North American context in the second decade of the twenty-first century.

Part 1: Perspectives

A significant conceptual area concerns the relationality between smart objects and how they embody and manifest 'intelligence'. Put differently, whatever definition we choose to adopt to describe intelligence, it will be a quality that emerges through the co-shaping relations between a smart object and its user. This first part of the book – Perspectives – offers ideas on fresh generative metaphors that can be used to think about and design smart objects as fungi, as actors in situated performances and as speculations on our hybrid and cyborgian futures by offering us different interpretations of machines. Rather than an intrinsic computational property or construed as a computing brain within the object, intelligence is enacted through multiple stabilities and relationalities. The work presented in this first part of the book explores alternative human-machine ontologies and perspectives on intelligence that make a practical contribution by helping designers envision, design and shape new morphologies for smart objects.

In Chapter 1, David Kirk, Effie Le Moignan and David Verweij examined how fungi, as living organisms, provide a powerful non-human metaphor for understanding smart objects. Interpreting smart objects by using fungal systems as an inspirational device allows conceptualizing them as hybrid entities part of, and generating, complex ecosystems of developing symbiotic relationships with human and non-human actors. Kirk and colleagues propose 'fungi' as a productive metaphor to imagine AI systems and ecologies of smart objects in a way that highlights slowness, otherness and coexistence. The chapter shows rather poetically how to look in unexpected places to generate new perspectives on functionality, application, human-AI partnerships and form factors. This perspective offers an alternative way of thinking about interaction with smart or intelligent interfaces, radically different from the usual anthropomorphic or zoomorphic metaphors.

Maaike Bleeker and Marco Rozendaal introduce the notion of a 'dramaturgy for devices' in Chapter 2, as a way to address interactions with smart objects as situated performances. In contemporary theatre, the term 'dramaturgy' refers to the totality of compositional principles that underpin the construction of performances. With their dramaturgy for devices, Bleeker and Rozendaal propose how smart objects can be understood not only through their technical computational properties but also through the relations they establish and transform within ecologies of people and things. As concrete suggestions for interaction design, they discuss how dramaturgical principles such as 'mise-en-scene', 'presence' and 'address' can help to guide designers to orchestrate such performances. Here the emphasis is on how to work with 'potentialities' and how through improvisations these potentialities might be actualized by means of design.

A satirical take on the imaginative potential of technology is present in the chapter that concludes this part. In Chapter 3, Tobias Revell and Kristina Andersen discuss how the notion of 'the machine' forges a cornerstone of our visions of the future. As humans, we dream and fear future machines as the true cyborgians we are; we fantasize through and with machines because they are more than simple tools and because we can imagine ourselves as one. The authors exhort us not to think in classifications (e.g. subject/object) but to remain open to the potential for new, evocative and alluring frameworks, to inform how objects and machines can be perceived and imagined. Put differently, the stories we tell each other about machines yet to exist tend to orient technological innovations. Likewise, we use innovations to forge new stories of futures that might (or might not) come to exist. Can 'better' machines be imagined, both in the quality of our imaginations and the machines therein? The chapter offers an answer by exploring speculative alternative machine ontologies.

Part 2: Interactions

The chapters in this part focus on how interactions between users and objects can be reimagined as an ongoing process of negotiation across multiple human and non-human actors, considering the multiplicity of identities, roles and embodiments they might assume. The same insight concerns the nature of agency as something that is distributed and emerges among networks and assemblages of people, objects and environments.

Notions of agency are introduced in Chapter 4 by Christopher Frauenberger, whose chapter draws on the work of Barad and Latour to develop a metaphysical position on the nature of the entanglements between humans and smart objects. To grasp the complexity of these relationships, the chapter argues, is necessary to portray them as a process of continuous negotiation for which appropriate spaces must be created and maintained, namely 'agonistic arenas' affording constructive conflicts over agency, power and morality with smart objects. To this aim, the chapter proposes the design of a new breed of smart objects: objects that are smart, honest and open to negotiate their relationships and material personalities with people around them, relationships that are co-developing by transforming the object in terms of its functionalities with the person's developing needs and interests.

In Chapter 5, Jelle van Dijk and Evert van Beek discuss the experience of smart objects from embodied and enactive perspectives, including the perspective of post-phenomenology. They illustrate how smart objects can display a dynamic kind of agency because of the multistable human-technology relations they can establish during interaction, that is, moving into the background or foreground of awareness and being perceived as tools or agents. Whether smart objects are seen as autonomous agents, or as social entities that we are in conversation with, often the underlying expectation is that these devices are 'in some sense like us'. By examining the ways in which smart objects 'can exist' as embodied agents in our everyday lived experience and 'dynamically mediate' human intentionality, the chapter offers insights on how their form and interactive behaviour can be designed accordingly. They conclude by suggesting design strategies that forefront ambiguity and openness to design for shifts in such emerging relationships.

To conclude this part, in Chapter 6 Nazli Cila and Carl DiSalvo conceptualize smart objects by examining delivery robots in the context of a smart city. They argue that ANT can help in critically revealing and explicating the actors and their qualities as sociotechnical contexts and networks, in which smart objects exist and operate. This perspective addresses objects and humans on a similar ontological level, thus with shared rights and responsibilities. Cila and DiSalvo propose that concepts from ANT can be mobilized by designers to help analyse and frame ecologies as expansive sociotechnical networks from the perspectives of all the actors involved, both human and non-human. This allows designers to 'see', identify and envision what is happening and hereby help them scope complex design spaces.

Part 3: Methodologies

This third part focuses on methods useful to the prototyping of smart objects and the form-giving typologies that may be exclusive to them. The expanding notion of smart objects as computational opens new ways to incorporate data and AI as a material to design with. It asks what kind of co-participatory methods will be needed in the near future, and examines how the design practice of prototyping can adapt to its particular emergent qualities and situatedness. We suggest that a research agenda should further interaction design practices where 'smartness' can be explored, questioned, sketched and prototyped. If it is true that interaction designers need to have a basic understanding of AI and ML, and of how these can be 'designed in' and incorporated into objects, it is also true that technical competence needs to be integrated by an understanding of ethics and the awareness that designing with data far too often reinforces existing social and cultural norms.

In Chapter 7, Philip van Allen explores how designers can approach the prototyping of smart things and contends that a paradigm shift in design practice is needed to adapt to their particular characteristics. Specifically, the craft of prototyping must adapt to new domains such as designing for unpredictability and emergence, contextual adaptation and animism, whilst accommodating key established design strategies – from sketching, rapid iteration, exploration and problem finding to user testing, participatory design and critical thinking. van Allen reviews some of the key techniques and methodologies to prototype with AI and ML, highlighting ways of carefully working with data sets that might have intrinsic biases and collaborating with data scientists as co-designers.

In Chapter 8, William Odom, Arne Berger and Dries De Roeck propose how co-design approaches can be used in practice to explore interactivity together with individuals and communities in a way that is embedded in the uniqueness and diversity of their everyday living situations. They also discuss involving people with different abilities and socially marginalized groups directly as stakeholders in the design of smart technologies. In this way, people's specific and highly individual circumstances (as social, material and political contexts) help shape smart objects and their complex ecologies in personally relevant ways. An obvious focus of intervention is the smart home. Their chapter questions the somewhat narrow conceptualizations of 'home' (what it is and how it is made) found in the fields of human-computer interaction and design. The chapter aims to expand such a vision of 'home', and the everyday domestic life it contains, by describing and critically reflecting on two design cases that offer different, yet complementary approaches, to the design of smart domestic technology, addressing participation and alternative lifestyles largely outside the mainstream.

Part 4: Critical Understandings

The final part of the book focuses on a critical understanding of smart objects in terms of their impact as social, legal and political entities. It proposes critical standpoints through which smart objects could be situated and politically theorized, as well as examined in the light of issues of responsibility, accountability and liability. These chapters offer salient reminders of the need to avoid common traps and tropes in the design of new technology. The good intentions of shaping a new technology's design are often overshadowed by the negative and unintended consequences that they give birth to. For interaction design to move forward, we need to better understand how to address power distribution (and asymmetries) within systems while safeguarding human integrity in their design and use. This concerns not simply the agency of things that are now able to act independently, making choices for us, but also the impact of this distributed agency on extended digital networks where 'objects' gather information, share this information about us and, crucially, communicate with each other outside of what is humanly perceivable.

In Chapter 9, Betti Marenko and Pim Haselager investigate technological fetishism and techno-determinism. The promise of technology as a sort of magical solution is still too pervasive among the privileged part of society, scholars, academics and technologists, and this kind of thinking is put to the test when actually bringing technology into society. Marenko and Haselager draw on a Marxist critique to address technology-induced alienation, techno-fetishism and life captured by an exploitative technocratic system that needs to keep on extracting people's data to function. Simple conveniences are traded for data, and the social structures of people's everyday lives can become regimented by smart objects. In a philosophical fiction, Marenko and Haselager imagine Karl Marx himself sitting on the sofa of a smart home intent in taking notes – exactly as he did in his analysis of the Industrial Revolution – and imagine how the world of smart objects would appear to him as an ecosystem of alienation-inducing commodities. The aim is to highlight pervasive deterministic assumptions concerning the role of digital technologies and impart a critical stance on the design of smart objects.

Chapter 10 by Ann Light questions the dominant narratives and agendas around smart technologies from the standpoint of her own first-person account of engaging with people in co-designing and planning for smart connected futures. Through the discussion of three research projects about future network technologies from the early 2000s to more recent times, Light critically examines the challenges in dealing with envisioning the invisible infrastructures of data and the difficulties of operating in contexts still in the making, fraught with indifference and scepticism. The observation of people's sense-making processes related to future technologies by what these innovations may tangibly afford them reveals key difficulties in how people conceptualize the notion of networks. Light advocates for values centred around the notions of care and empathy to be embedded in the design of smart objects by enabling collective participation to make a bridge between academic research and people's everyday concerns, which ultimately facilitates interconnectedness with each other and our planet.

A consideration of the ethical, legal and societal implications (ELSI) of AI is discussed by Pim Haselager in Chapter 11. His chapter is a reminder that while smart objects ought, ideally, add their own smartness to that of their users so as to improve overall functionality and experience, in practice, however, such mixes of human and non-human intelligences might lead to unfavourable and unpredictable outcomes, and to increased risk of undesirable consequences. Worst of all, the use of smart objects might lead to users' uncertainty about agency, responsibility and liability, and a lack of clarity about who, or what, is in charge. Haselager makes a plea for the development of 'wise objects': smart objects that adhere to ethical, legal and societal constraints and minimize agency and responsibility confusions. This shift from 'smart' to 'wise' further opens up an imaginary space that, by envisioning smart objects that can go against the requests or actions of their users, that are responsible and that 'know when to quit', can inspire designers in prototyping increasingly protective, reliable and trustworthy smart objects.

A launch into the future

This book should be taken both as a snapshot of the present situation as well as an indication of the terms of a future research agenda, which we argue is transdisciplinary, process-oriented and relational. The research agenda that this book puts forward offers practical suggestions through design speculations, interventions and practices, aims to participate in future societal transformations and triggers reflection, dialogue and debate. This research agenda wants to enable the design and development of smart objects within technological and commercially driven environments and industries, while providing a robust critique to sustain such development. One thing is clear, a research agenda for interaction design demands practices, modes of thinking and ethical standpoints, as well as new vocabularies and images to think with. It centres on an understanding and design of smart objects that embrace their hybrid nature as shifting and blending tools, agents, machines and even 'creatures' that can enter into multiple kinds of relationships with us humans that are meaningful and empowering in the context of everyday life. It aims to illuminate hidden infrastructures behind the functioning of smart objects by stirring debates centring on technology, human values and impact on economy and ecology. We hope that reading this book will provide you, the reader, with inspiration on how to engage in this agenda as a scholar, design practitioner or activist. Finally, we want to hear from you on how these ideas resonate with your own practices in academia, industry or education, and engage in a dialogue that we hope can start as the book ends.

Bibliography

Aarts, E., & Wichert, R. (2009). Ambient intelligence. In *Technology guide* (pp. 244–249). Berlin: Springer.

Ackermann, E. (2005). Playthings that do things: A young kid's 'incredibles'! In *Proceedings of the 2005 Conference on Interaction Design and Children (IDC '05)* (pp. 1–8). New York: ACM.

Al-Fuqaha, A., Guizani, M., Mohammadi, M., Aledhari, M., & Ayyash, M. (2015). Internet of things: A survey on enabling technologies, protocols, and applications. *IEEE Communications Surveys & Tutorials*, 17(4), 2347–2376.

Allen, J. F., Byron, D. K., Dzikovska, M., Ferguson, G., Galescu, L., & Stent, A. (2001). Toward conversational human-computer interaction. *AI Magazine*, *22*(4), 27.

Appadurai, A. (Ed.). (1986). *The social life of things: Commodities in cultural perspective.* Cambridge: Cambridge University Press.

Atzori, L., Iera, A., & Morabito, G. (2010). The internet of things: A survey. *Computer Networks*, *54*(15), 2787–2805.

Auger, J. (2014). Living with robots: A speculative design approach. *Journal of Human-Robot Interaction*, *3*(1), 20–42.

Barad, K. (2007). *Meeting the universe half-way: Quantum physics and the entanglement of matter and meaning.* Durham: Duke University Press.

Bødker, S., & Andersen, P. B. (2005). Complex mediation. *Human-computer interaction*, *20*(4), 353–402.

Boer, L., & Bewley, H. (2018, June). Reconfiguring the appearance and expression of social robots by acknowledging their otherness. In *Proceedings of the 2018 on Designing Interactive Systems Conference 2018* (pp. 667–677). New York: ACM.

Braitenberg, V. (1986). *Vehicles: Experiments in synthetic psychology.* Cambridge, MA: MIT Press.

Breazeal, C. (2003). Toward sociable robots. *Robotics and autonomous systems, 42*(3–4), 167–175.
Brooks, R. A. (1991). Intelligence without representation. *Artificial intelligence, 47*(1–3), 139–159.
Candlin, F., & Guins, R. (2008). *The object reader*. Oxford: Routledge.
Chamberlain, A., Crabtree, A., Rodden, T., Jones, M., & Rogers, Y. (2012, June). Research in the wild: Understanding 'in the wild' approaches to design and development. In *Proceedings of the Designing Interactive Systems Conference* (pp. 795–796). New York: ACM.
Cila, N., Smit, I., Giaccardi, E., & Kröse, B. (2017, May). Products as agents: Metaphors for designing the products of the IoT age. In *Proceedings of the 2017 CHI Conference on Human Factors in Computing Systems* (pp. 448–459). New York: ACM.
Darwin, C. (2004). *On the origin of species, 1859*. Oxford: Routledge.
D'Adderio, L. (2011). Artifacts at the centre of routines: Performing the material turn in routines theory. *Journal of Institutional Economics*, 7(2), 197–230.
De Certeau, M. (2011). *The practice of everyday life*. Berkeley: University of California Press.
Dennett, D. C. (1989). *The intentional stance*. Cambridge, MA: MIT Press.
Desjardins, A, Viny, J., Key, C. and Johnston, N. (2019). Alternative avenues for IoT: Designing with non-strereotypical homes. In *Proceedings of the 2019 CHI Conference on Human Factors in Computing Systems*, paper 351 (pp. 1–13). New York: ACM.
Dourish, P. (2004). *Where the action is: The foundations of embodied interaction*. Cambridge, MA: MIT Press.
Dove, G., Halskov, K., Forlizzi, J., & Zimmerman, J. (2017, May). UX design innovation: Challenges for working with machine learning as a design material. In *Proceedings of the 2017 Chi Conference on Human Factors in Computing Systems* (pp. 278–288). New York: ACM.
Engemann, C., & Feigelfeld, P. (2017) Distributed embodiment. In *Hello, Robot! Design between human and machine*. Exhibition catalogue (pp. 252–259). Weil am Rhein, Germany: Vitra Design Museum GmbH.
Engeström, Y. (1999). Activity theory and individual and social transformation. *Perspectives on Activity Theory*, 19(38), 19–30.
Farooq, U., & Grudin, J. (2016). Human-computer integration. *Interactions, 23*(6), 26–32.
Fogg, B. J. (2003) *Persuasive technology: Using computers to change what we think and do*. San Francisco: Morgan Kaufmann.
Følstad, A., & Brandtzæg, P. B. (2017). Chatbots and the new world of HCI. *Interactions, 24*(4), 38–42.
Forlizzi, J. (2008). The product ecology: Understanding social product use and supporting design culture. *International Journal of Design, 2*(1), 11–20.
Forlizzi, J., & DiSalvo, C. (2006, March). Service robots in the domestic environment: A study of the Roomba vacuum in the home. In *Proceedings of the 1st ACM SIGCHI/SIGART Conference on Human-Robot Interaction* (pp. 258–265). https://doi.org/10.1145/1121241.1121286.
Fromm, J. (2005). Types and forms of emergence. *arXiv*. Retrieved 21 March 2021 from https://arxiv.org/abs/nlin/0506028.
Funk, M., Eggen, B., & Hsu, J. Y. J. (2018). Designing for systems of smart things. *International Journal of Design, 12*(1), 1–5.
Giaccardi, E., Speed, C., Cila, N., & Caldwell, M. (2016). Things as co-ethnographers: Implications of a thing perspective for design and anthropology. In R. C. Smith, K. T. Vangkilde, M. G. Kjaersgaard, T. Otto, J. Halse & T. Binder (Eds), *Design anthropological futures* (pp. 235–248). Oxford: Routledge.

Gibson, J. J. (1979). *The ecological approach to visual perception: Classic edition*. Hove, East Sussex, UK: Psychology Press.
Govers, P., Hekkert, P., & Schoormans, J. P. (2003). Happy, cute and tough: Can designers create a product personality that consumers understand. In D. McDonagh, P. Hekkert, J. van Erp & D. Gyi (Eds), *Design and emotion, Episode III: The experience of everyday things* (pp. 345–349). London: Taylor & Francis.
Gray, K., & Wegner, D. M. (2012). Feeling robots and human zombies: Mind perception and the uncanny valley. *Cognition*, 125(1), 125–130.
Greenfield, A. (2006). *Everyware: The dawning age of ubiquitous computing*. San Francisco, CA: New Riders.
Greenfield, A. (2017). *Radical technologies: The design of everyday life*. London: Verso.
Greengard, S. (2015). *The internet of things*. Cambridge, MA: MIT Press.
Hearst, M. A., Allen, J., Guinn, C., & Horvitz, E. (1999). Mixed-initiative interaction: Trends and controversies. *IEEE Intelligent Systems*, *14*(5), 14–23.
Heider, F., & Simmel, M. (1944). An experimental study of apparent behavior. *The American Journal of Psychology*, *57*(2), 243–259.
Hoffman, G., Kubat, R., & Breazeal, C. (2008, August). A hybrid control system for puppeteering a live robotic stage actor. In *The 17th IEEE International Symposium on Robot and Human Interactive Communication*, RO-MAN 2008 (pp. 354–359). New York: IEEE.
Hoffman, G., & Ju, W. (2014). Designing robots with movement in mind. *Journal of Human-Robot Interaction*, *3*(1), 89–122.
Holmquist, L. E. (2017). Intelligence on tap: AI as a new design material. *Interactions*, *24*(4), 28–33.
Horvitz, E. (1999, May). Principles of mixed-initiative user interfaces. In *Proceedings of the SIGCHI Conference on Human Factors in Computing Systems* (pp. 159–166). New York: ACM.
Hutchins, E. (1995). *Cognition in the Wild*. Cambridge, MA: MIT Press.
Ihde, D. (1990). *Technology and the lifeworld: From garden to earth* (No. 560). Bloomington: Indiana University Press.
Jacucci, G., Spagnolli, A., Freeman, J., & Gamberini, L. (2014, October). Symbiotic interaction: A critical definition and comparison to other human-computer paradigms. In *International Workshop on Symbiotic Interaction* (pp. 3–20). Cham: Springer.
Janlert, L. E., & Stolterman, E. (1997). The character of things. *Design Studies*, *18*(3), 297–314.
Jung, H., Stolterman, E., Ryan, W., Thompson, T., & Siegel, M. (2008). Toward a framework for ecologies of artifacts: How are digital artifacts interconnected within a personal life? In *Proceedings of the 5th Nordic Conference on Human-Computer Interaction: Building Bridges* (pp. 201–210). New York: ACM.
Kaptelinin, V., & Nardi, B.A. (2006). *Acting with technology: Activity theory and interaction design*. Cambridge, MA: MIT Press.
Kortuem, G., Kawsar, F., Sundramoorthy, V., & Fitton, D. (2010). Smart objects as building blocks for the internet of things. *IEEE Internet Computing*, 14(1), 44–51.
Kuijer, L., Jong, A. D., & Eijk, D. V. (2013). Practices as a unit of design: An exploration of theoretical guidelines in a study on bathing. *ACM Transactions on Computer-Human Interaction (TOCHI)*, *20*(4), 21.
Kuijer, L., & Giaccardi, E. (2018, April). Co-performance: Conceptualizing the role of artificial agency in the design of everyday life. In *Proceedings of the 2018 CHI Conference on Human Factors in Computing Systems* (p. 125). New York: ACM.
Kuniavsky, M. (2010). *Smart things: Ubiquitous computing user experience design*. Amsterdam: Elsevier.

Kuutti, K., & Bannon, L. J. (2014, April). The turn to practice in HCI: Towards a research agenda. In *Proceedings of the 32nd Annual ACM Conference on Human Factors in Computing Systems* (pp. 3543–3552). New York: ACM.
Lakoff, G., & Johnson, M. (1980). Conceptual metaphor in everyday language. *The Journal of Philosophy, 77*(8), 453–486.
Latour, B. (1993). *We have never been modern.* Cambridge, MA: Harvard University Press.
Latour, B. (2005). *Reassembling the social: An introduction to actor-network-theory.* Oxford: Oxford University Press.
Laurel, B. (1997). Interface agents: Metaphors with character. In Batya Friedman (Ed.), *Human Values and the Design of Computer Technology* (pp. 207–219). Stanford: CSLI Publications.
Leontiev, A. N. (1975). *Activities. Consciousness. Personality.* Moscow: Politizdat.
Levillain, F., & Zibetti, E. (2017). Behavioral objects: The rise of the evocative machines. *Journal of Human-Robot Interaction, 6*(1), 4–24.
Marenko, B. (2014). Neo-animism and design: A new paradigm in object theory. *Design and Culture, 6*(2), 219–241.
Marenko, B., & van Allen, P. (2016). Animistic design: How to reimagine digital interaction between the human and the nonhuman. *Digital Creativity, 27*(1), 52–70.
Mataric, M. J. (1993, April). Designing emergent behaviors: From local interactions to collective intelligence. In *Proceedings of the Second International Conference on Simulation of Adaptive Behavior* (pp. 432–441). https://doi.org/10.7551/mitpress/3116.003.0059.
McTear, M. F., Callejas, Z., & Griol, D. (2016). *The conversational interface.* Cham: Springer.
Michalski, R. S., Carbonell, J. G., & Mitchell, T. M. (Eds). (2013). *Machine learning: An artificial intelligence approach.* New York: Springer Science & Business Media.
Mori, M., MacDorman, K. F., & Kageki, N. (2012). The uncanny valley [from the field]. *IEEE Robotics & Automation Magazine, 19*(2), 98–100.
Mulgan, G. (2018) *Big mind: How collective intelligence can change our world.* Princeton, NJ: Princeton University Press.
Nass, C., Steuer, J., Henriksen, L., & Dryer, D. C. (1994). Machines, social attributions, and ethopoeia: Performance assessments of computers subsequent to. *International Journal of Human-Computer Studies, 40*(3), 543–559.
Norman, D. (1993). *Things that make us smart: Defending human attributes in the age of the machine.* New York: Basic Books.
Odom, W., & Duel, T. (2018, April). On the design of OLO radio: Investigating metadata as a design material. In *Proceedings of the 2018 CHI Conference on Human Factors in Computing Systems* (p. 104). New York: ACM.
Odom, W., Zimmerman, J., Davidoff, L. S., Forlizzi, J., Dey, A. K., & Lee, M. K. (2012). A fieldwork of the future with user enactments. In *Proceedings of DIS 2012* (pp. 338–347). New York: ACM.
Odom, W. T., Sellen, A. J., Banks, R., Kirk, D. S., Regan, T., Selby, M., *et al.* (2014, April). Designing for slowness, anticipation and re-visitation: A long term field study of the photobox. In *Proceedings of the SIGCHI Conference on Human Factors in Computing Systems* (pp. 1961–1970). New York: ACM.
Oogjes, D., Odom, W., & Fung, P. (2018, June). Designing for an other home: Expanding and speculating on different forms of domestic life. In *Proceedings of the 2018 on Designing Interactive Systems Conference 2018* (pp. 313–326). New York: ACM.
Pierce, J., Strengers, Y., Sengers, P., & Bødker, S. (2013). Introduction to the special issue on practice-oriented approaches to sustainable HCI. *ACM Transactions on Computer-Human Interaction (TOCHI)*, 20(4), 1–8.

Redström, J., & Wiltse, H. (2018). *Changing things: The future of objects in a digital world.* London: Bloomsbury.
Reeves, B., & Nass, C. I. (1996). *The media equation: How people treat computers, television, and new media like real people and places.* Cambridge: Cambridge University Press.
Robles, E., & Wiberg, M. (2010, January). Texturing the material turn in interaction design. In *Proceedings of the fourth international conference on Tangible, embedded, and embodied interaction* (pp. 137–144). New York: ACM.
Rose, D. (2014). *Enchanted objects: Design, human desire, and the internet of things.* New York: Simon & Schuster.
Rozendaal, M. C. (2016). Objects with intent: A new paradigm for interaction design. *Interactions, 23*(3), 62–65.
Rozendaal, M. C., Ghajargar, M., Pasman, G., & Wiberg, M. (2018). Giving form to smart objects: Exploring intelligence as an interaction design material. In Michael Filimowicz and Veronika Tzankova (Eds), *New Directions in Third Wave Human-Computer Interaction: Volume 1-Technologies* (pp. 25–42). Cham: Springer.
Rubinshtein, S. L. (1946). *Foundations of general psychology.* Moscow: Academic Science.
Scholl, B. J., & Gao, T. (2013). Perceiving animacy and intentionality: Visual processing or higher-level judgment. In M. D. Rutherford and V. A. Kuhlmeier (Eds), *Social perception: Detection and interpretation of animacy, agency, and intention* (pp. 197–229). Cambridge, MA: MIT Press.
Shove, E. (2007). *The design of everyday life.* Oxford: Berg.
Sterling, B. (2005). *Shaping Things – mediawork pamphlets.* Cambridge, MA: MIT Press.
Stolterman, E., & Croon Fors, A. (2008). Critical HCI research: A research position proposal. *Design Philosophy Papers*, 1, 23.
Suchman, L. (2007). *Human-machine reconfigurations: Plans and situated actions.* Cambridge: Cambridge University Press.
Suchman, L. (2011). Subject objects. *Feminist Theory, 12*(2), 119–145.
Takayama, L. (2009, March). Making sense of agentic objects and teleoperation: In-the-moment and reflective perspectives. In *2009 4th ACM/IEEE International Conference on Human-Robot Interaction (HRI)* (pp. 239–240). New York: IEEE.
Taylor, A. S. (2009, April). Machine intelligence. In *Proceedings of the SIGCHI Conference on Human Factors in Computing Systems* (pp. 2109–2118). New York: ACM.
Turkle, S. (2007). *Evocative objects: Things we think with.* Cambridge, MA: MIT Press.
Vallgårda A., (2014) Giving form to computational things: Developing a practice of interaction design. *Personal Ubiquitous Computing, 18*(3), 577–592.
Verbeek, P-P. (2005). *What things do: Philosophical reflections on technology, agency, and design.* State College: Penn State Press.
Vygotsky, L. S. (1978). *Mind in society: The development of higher mental process.* Cambridge, MA: Harvard University Press.
Wakkary, R., Oogjes, D., Hauser, S., Lin, H. W., Cao, C., Ma, L., *et al.* (2017, June). Morse Things: A design inquiry into the gap between things and us. In *Proceedings of the 2017 Conference on Designing Interactive Systems (DIS '17)* (pp. 503–514). New York: ACM.
Wakkary, R., Oogjes, D., Lin, H. W., & Hauser, S. (2018, April). Philosophers living with the Tilting Bowl. In *Proceedings of the 2018 CHI Conference on Human Factors in Computing Systems*, paper 94 (pp. 1–12). New York: ACM.
Weiser, M. (1991). The computer for the 21st century. *Scientific American, 265*(3), 94–104.
Wiberg, M. (2014). Methodology for materiality: Interaction design research through a material lens. *Personal and Ubiquitous Computing, 18*(3), 625–636.
Wooldridge, M., & Jennings, N. R. (1995). Intelligent agents: Theory and practice. *The Knowledge Engineering Review, 10*(2), 115–152.

PART ONE

PERSPECTIVES

1 AN ILLUSTRATED FIELD GUIDE TO FUNGAL AI FOR DESIGNERS

David Kirk, Effie Le Moignan and David Verweij

When we are designing digital technologies, we are often inspired by patterns we see elsewhere and seek to replicate them, as metaphors and analogies, in an effort to make the interfaces to those technologies easier to use or more understandable for the user (Norman, 2013). This is as true for interaction design as it is for product design, two fields which are increasingly interwoven (Rowland, Goodman, Charlier, Light & Liu, 2015). We are moving towards a world of smart domestic products and devices, networked artefacts that connect both to us and to one another, imbued with the vestiges of 'intelligence'. Consequently, it is likely that we will continue to need to utilize metaphor and analogy to make these devices intelligible to their users (*ibid.*).

Within interaction design, the discussion of the role of metaphor and analogy at the interface has formed an extensive part of an ongoing debate about the nature of what some term 'Natural User Interfaces' or Reality-Based Interaction (Jacob *et al.*, 2008). However, as we think through how future intelligent devices will interact with us, we might question exactly how notions of intelligence are constructed. Where do we find the examples of intelligence that we draw upon, for our inspiration and our metaphors, and what are the assumptions and implications hidden within these metaphors?

Webster and Weber (2007) highlight the characteristics of fungi which differentiate them from animal and plant life. As stationary forms sited on a substrate host material, fungi feed by absorption of nutrients from the environment.

The enzymes produced by fungi render nutrients available by breaking down the cells of the host material, for absorption. This leads to a wide range of fungal forms suitable for diverse environmental conditions, from single-celled yeasts to complex mushrooms. These range in ecology from those which are freeform, to those mutually symbiotic with a host, to the parasitic and hyperparasitic (*ibid.*). As a result, fungi are a ubiquitous presence in terrestrial and freshwater environments (Dix and Webster, 1995), with a massive number observed and predicted to exist, ranging from a conservative 1.5 million species to 9.9 million species globally (Hawksworth, 2001). Fungi form the largest biomass in soil, act as decomposition agents for dead materials and pathogens, make nutrients bioavailable to plants and have roles in the nitrogen and carbon cycles (Gadd, Watkinson & Dyer, 2007). Fungi produce spores which facilitate their spread, but additionally can form substantial mycelium networks which penetrate the soil, forming extensive networks (Webster & Weber, 2007). These networks are symbiotic with plants, acting as chemical communication networks signalling to a plant when another nearby is under attack from aphids (Babikova *et al.*, 2013). Fungi are complicated, multiform and act in ways which serve to interact and respond to their environment that are deeply and fundamentally different from those in which humans or animals do.

Using the diverse and unusual characteristics found within fungi, we utilize them within this chapter as a metaphor for exploring AI in everyday settings, focusing on the home (as a typical site for encountering 'smart objects'). Presented as a brief illustrated guide to seven conceptual forms, we introduce the notion of 'Fungal AI' as a provocation for design, exploring the idea of drawing on mycology (the study of mushrooms and fungus) as a novel source of inspiration for the design of smart objects. Fungi have long been utilized for a range of practical purposes, from baking and brewing with yeasts, as foodstuffs, for medicinal properties and for religious and cultural use as hallucinogens (Boa, 2004; Milenge Kamalebo, Malale, Ndabaga, Degreef & De Kesel, 2018).

In this chapter we begin to explore the characteristics of fungi as a starting point of speculation for future design, in contrast to more commonly employed anthropomorphic and zoomorphic approaches to the design of interfaces, and look at the play of possibilities (Anderson, 1994) when using alternative conceptions of intelligence.

Useful properties of fungi or mushrooms

Fungi, quite broadly, offer some potential properties, which we might utilize as primary design characteristics for inspiration and which may mark them out as distinct from other kinds of intelligence that we routinely encounter.

With a concept such as a fungal AI, there might be two possible interpretations – the first is that we are advocating some kind of complex bioengineering in which fungal DNA is very literally made synthetic and algorithmic in a way where we can very literally create new forms of life. The second idea we are advocating is that, as a design community, we might take inspiration from fungal properties, through metaphor, to explore new kinds of AI-like systems, new smart services and technologies, for domestic life (for example) that draw on the kinds of properties that can be found in the fungal world. It is this latter interpretation of the concept of fungal AI that we are hoping to pursue herein.

Whilst fundamentally complicated in terms of the biological structures and chemical processes which occur in some fungi, there are key points which render them as suitable for consideration in terms of AI systems. Broadly, one of these is that fungi require a substrate base on which to proliferate and to provide access to resources needed to thrive. This makes them ideally suited to system design concepts of embedding, input and output, and, for example, working with the everyday residua of smart objects such as data traces and network traffic. Fungi respond to their environment, monitoring and adapting to conditions, thereby showing learnt behaviours akin to the machine learning of 'smart objects'. It is worth noting that these characteristics, which might underpin the metaphor, are readily understood by the non-expert. It is of note generally that fungi

- are orthogonal to human desires and intentions and exist in their own right (have their own needs, goals, and so forth)
- make no attempt to interact directly with us but provide functional value to humans, animals and plants
- grow, learn and adapt to different ecologies
- propagate themselves (sometimes with support of other organisms)
- this action can be facilitated by providing appropriate 'sites of propagation'
- coexist with humans in our spaces (and on/in our bodies)

As fungi do not communicate or appear as directly responsive, they represent a departure from other forms of life. It is relatively easy to ascribe intelligence, motive or characteristics onto animals, for example. This is due to similarity in form, such as having a face, and overlapping needs for food, warmth and shelter. We can recognize playfulness, sociability and states of being. However, it is more challenging to speculate around intelligence which is divorced from embodied forms and states we can identify with – or indeed easily project an illusion of familiarity upon (whether this is accurate or not). Fungi are intensely responsive to their environment (albeit at temporal scales which feel alien to us or are hard for us to perceive), using it as the very material on which to grow and derive their nutrients from. Whilst some forms exist in symbiotic or parasitic forms, they are generally self-contained and independent. There is no 'window into the soul' to interpret or interact with a fungal form. One can provoke a response by changing the environment, which is observable. It can be viewed as naturally spreading and encouraged to grow by propagation. However, the needs of fungi are functional, and their existence is very alien to us, despite the fact that we often share our homes (and even our bodies) with these 'intelligent' life forms.

Ultimately, the benefit of fungal AI systems to people might be quite tangential – we might design/cultivate a fungal AI and find that it is performing some kind of function or offering some kind of service that we can farm – that we can take advantage of whilst it is in essence 'doing its own thing'.

Spotting common fungal AI in the wild

In this section we provide an alternative design fiction in the form of a set of preliminary written field notes on different kinds of AI mushroom morphology and how one might encounter them in the wild – each of which shows a fungus-like AI working in a different way – each offering different tangential functionality to humans. We situate these fungal AIs (illustrated in Figure 1.1) primarily in

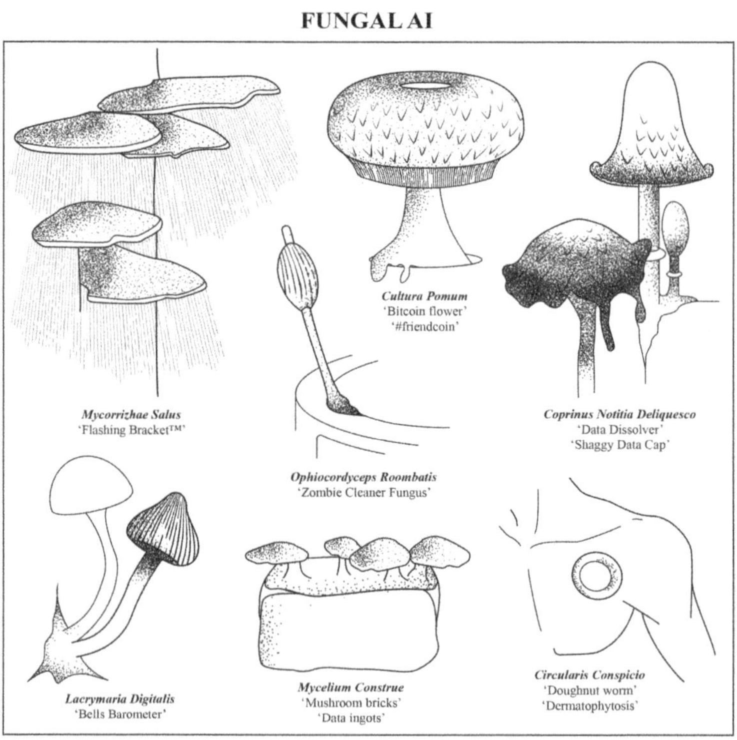

FIGURE 1.1 Fungal AI.

the domestic space, to note the common way in which we live alongside these often unremarkable but useful life forms and because they speak to the prevalent locations in which we might encounter smart objects. This allows us to begin to demonstrate the utility of the metaphor and to explore some of the play of possibilities engendered by the concept of fungal AI.

Lacrymaria digitalis

Bells barometer

[**Distribution**] Large areas of Europe and North America and most temperate zones with inclement weather, particularly common in the north-east of England

[**Habitat**] Grows/propagates near damp and cool surfaces such as windows where there is a clear line of sight to external precipitation and audible sound of rainfall

[**Properties/behaviours**] Estimates rainfall for central data collation by measuring onset and offset of rain sounds. Will learn to bloom with the imminent onset of precipitation. Can be used to provide rainfall data for passing apps.

[**Visual description**] Pinkish in colour, with light brown gills, fruiting body can tend towards the tear shape

Coprinus notitia deliquesco

Data dissolver; Shaggy data cap

[**Distribution**] Whole of Europe or any General Data Protection Regulation (GDPR)-compliant zone

[**Habitat**] Likely to be found growing on or around wireless-connected security systems in domestic environments

[**Properties/behaviours**] Grows in size as it aggregates domestic movement and activity data from nearby building/security sensors. Learns and archives patterns of movement for specific occupants. Using digital mycorrhizal connections, fruiting body data will be exchanged with smart thermostats to dynamically correlate heating/cooling requirements with occupancy and activity measures. When building occupants have left for an extended period of days, the mushroom will dissolve itself in an autodigestion process, purging personally identifiable data and leaving behind a pool of data residua, which serves as an algorithmic growth medium for new blooms or can be archived as an historic resource of privacy-preserved building data.

[Visual description] Tall white caps, somewhat scaly. The gills beneath are white when young but then turn inky black and dissolve when data is decomposing.

Ophiocordyceps roombatis

Zombie cleaner fungus

[Distribution] Wide geographic spread – both northern and southern hemispheres, mostly in industrialized areas, particularly common in 'smart homes'

[Habitat] An iotopathogenic fungus currently found predominantly in smart homes, specifically infecting models of the iRobot Roomba Robotic Vacuum species

[Properties/behaviours] Full pathogenesis will be enacted by the host device (Roomba) having its behaviour artificially manipulated. The fungal AI influences and reprograms intelligent circuitry; in particular, infected Roombas will have their navigation altered. By manipulating the Roomba's smart dock, and then altering movement repertoires, the fungus steers Roombas in to 'foraging trails' to identify networked smart home hubs (such as your Amazon Alexa). Wireless signal strength is then mapped around the house, and through manipulating Bluetooth connectivity, the fungus learns to identify smart objects in the home (as opposed to nearby houses) and will automatically configure home networks to add new devices whilst the Roomba cleans.

The fungus is known to engage in an active secondary metabolism producing anti-malware agents as part of a fungus-host ecosystem, which benefits the robot. When a Roomba eventually reaches the end of its service life, an emerging fruiting body erupts from the 'head' of the Roomba. When the Roomba is left next to an uninfected (new) cleaning device, 'data spores' will be released; this propagates current models of network configurations. The fungus may also parasitize other similarly programmed and configured species of robotic vacuum, but likely with lesser degrees of host manipulation and reproductive success.

[Visual description] Only visible at end of vacuum working life. Wiry tendrils emerge from nape of photocell sensor. Upright structures support a darkly pigmented vesicle holding the spore-bearing sexual structures.

Mycorrizhae salus

Flashing Bracket™

[Distribution] Currently only available in Japan

[Habitat] Exterior surfaces of domestic buildings

[Properties/behaviours] Must be attached to the outside of the house – as it requires line of sight to nearby 'growths' to work with full community effect. It is wirelessly networked to the devices inside the house using the mycorrhizal fungal AI network. The network monitors data traffic patterns between devices in the home. It is particularly sensitive to infections on the network and will respond to unusual traffic flows around specific devices by alerting connected devices to the unusual network device activity. When infection is detected, the fungus's exterior fruiting body glows in a light spectrum visible to other fungi on other buildings. Light-pulsing patterns share information about the nature of the attack with other fungal installations on other buildings, requiring line of sight – similar to a 5G network – offering non-IP network communication of security threats. It allows networks of community buildings to prepare for and prevent certain kinds of network infiltration attacks from propagating. Bioluminescent properties can also provide useful low-key exterior security lighting.

[Visual description] Shelf-like hemispherical brackets of varying colours, ranging in size from 4- to 30-cm spans, and exhibits bioluminescence at night

Circularis conspicio ('circular marks')

Doughnut worm; dermatophytosis

[Distribution] Worldwide with the exclusion of exceptionally cold climates (where vasoconstriction limits blood supply to the epidermis, restricting conditions to thrive)

[Habitat] Human upper epidermis. Upper torso and radial sections of limbs

[Properties/behaviours] Fungus appears in ring-like markings as a bio-response to the host as growth medium. Appears on the human body as transitory characteristic ring-like markings in response to bodily fluctuations in homeostatic conditions. The rings visibility alters in response to bodily levels of blood sugar, providing an indicative measure of the host's glycaemic state. Raised welt-like circular patches, with a varying opacity and prominence of a pinkish undertone

[Visual description] Tightly formed, rash-like clusters

Cultura pomum

Bitcoin flower; #friendcoin

[**Distribution**] Worldwide, contained to communities with digital data-sharing capacities (which do not prohibit the sale of data in the mycelium intermediary treated format)

[**Habitat**] Communal domestic spaces

[**Properties/behaviours**] Via the collection of household data within the home from multiple devices, the fungi performs telemetry to filter the data, anonymizing it and providing it to external parties who may wish to access data on households. This symbiosis encourages prosocial behaviours within the home (recycling, energy efficiency, sharing resources) by rewarding the occupants with virtual currency when it fruits, in return for data stream access. Thus, *Cultura pomum* requires access to homes in order to produce fruit, and rewards the occupants, which incentivises access. This domestic currency exchange is mutually beneficial for both parties.

[**Visual description**] Long pale body, with darker wrinkled and pitted head; gills emerge as a lace-like skirt, from the base of the head, when fruiting, with the structure revealing a unique machine-readable pattern, which fades after virtual currency is harvested.

Mycelium construe (fungal building/constructing)

Mushroom bricks; data ingots

[**Distribution**] Europe and North America

[**Habitat**] Warm moist spaces with access to 3D printers

[**Properties/behaviours**] The fungus will propagate, typically as spores moved between sites, on LANs with access to 3D print functionality and local networked sensors. These mycelia form 'bricks' in response to data collection from the home. In response to stimuli, such as auditory inputs, the machine learning generates new data points and networks which produce dense, tightly packed informational structures. These provide a light, dense and dry overall formation to the bricks. Those bricks produced using differing input data may result in bricks of differing density and weight depending on the AI. This generative process is continual, building a series of bricks from home data over time, which are slowly extruded from connected 3D print facilities.

[**Visual description**] Dense formations of extremely small beige or brown individual fungus in fibrous-looking extrusions which collectively build an irregular mass that may collapse if left to accrue organically; within confines these structures it will take the form of a mould or container.

Fungi as an alternative metaphor

What does fungi have to offer, as an alternative metaphor to the more common anthropomorphic and zoomorphic forms encountered in design? Part of their potential lies in the departure from forms with which we identify, or attempt to identify with, and sense-make as familiar. The otherly nature of fungi (as something with which society engages, communicates or innately recognizes as "not" living or "not" responsive to us) disrupts deep-seated patterns in design. Very often the presence of mould is in fact unwelcome, as seen with common fungi in mildew or spoiled food.

For several generations we have explored the anthropomorphization (making things humanlike in form) of proposed, and increasingly realized, smart devices such as robots (Dereshev, 2018). They are commonly given human bodily forms, faces with attendant eyes and ears for coordinating attention, and limbs for grasping, articulation and gestural support (Breazeal, 2002). Even when intelligence is not intended to be embodied, and is designed to be purely ambient, interaction designers and engineers have increasingly looked to support the capabilities of conversational interaction, which again imply, and trade on, certain kinds of humanlike (anthropomorphic) intelligence. Classic sci-fi interpretations of ambient intelligence in 'domestic' spaces have extensively explored these tropes. For example, 'Hal' in Stanley Kubrick's *2001: A Space Odyssey* is a well-understood and commonly referenced dystopian vision of the conversational ambient intelligence.

Anthropomorphization, however, is not without problems. Adopting both humanlike physical forms and interaction modalities (such as voice-based interaction) can lead to interactional difficulties. Two commonly cited issues associated with this are the over-assumption of intelligent capabilities of a device, wherein users assume that devices are smarter than they really are (Nowacka, Hammerla, Elsden, Plötz & Kirk, 2015) and the 'uncanny valley' (Mori, MacDerman & Kageki, 2012). Here users are viscerally repulsed to varying degrees by the unnaturalness of an otherwise nearly natural appearing object. It is also very common to hear ethical objections to the use of unmitigated anthropomorphization of intelligent interactions. This was perfectly illustrated when Google's voice-based 'Duplex' AI system was revealed at a press conference and was demonstrated pretending to be a human caller. This inserted plausible, breath and prevarication pauses to make the dialogue sound more humanlike, whilst making appointments with local businesses (Griffin, 2018).

Across many fields of design there has been exploration of biomimicry (see https://www.dezeen.com/tag/biomimicry/) as a source for inspiration.

Within interaction design, and in particular with application to 'human-robot interaction', there has been extensive use of not only anthropomorphization but also zoomorphization (exploring form factors for smart interactive devices which borrow heavily from analogy to animal forms and animal-like intelligences). Well-known examples include military technologies such as Boston Dynamics 'Big Dog' (Boston Dynamics, 2004), toys such as Pleo and Furby (Fernaeus, Håkansson, Jacobsson & Ljungblad, 2010) and even robotic devices such as autonomous vacuum cleaners (Forlizzi & DiSalvo, 2006). The case of Roomba, a robotic vacuum cleaner, is of particular interest here, as several studies have explored what happens when people come to live with these kinds of devices. Studies such as those by Ja Sung, Rebecca Grinter and Henrik Christensen (2009) and Florian Vaussard *et al.* (2014) have explored what it means 'to make such a device at home' in the domestic environment. These works have shown the ways in which even if the technology itself has a hard and appliance-like form factor, it will be oriented to as if an animal is within the home, with animal-like intelligent 'qualities'. Users will commonly impute certain kinds of intentionality and agency to the device, which instantly imply that they are not treating it as an inorganic, algorithmically driven piece of hardware, executing preprogrammed instructions (which it is). Users have been observed softening, furring and otherwise making these devices more visually animal-like, and have extensively documented online their interactions with pet-devices' in the home (http://tiny.cc/atdsgz). This demonstrates the extent to which users need to find a suitable framing for an object with apparent agency operating within their homes to make themselves more comfortable with the device, and evidently the framing of 'pet' seems readily at hand. Arguably, this also demonstrates a kind of innate human – but possibly mammalian – tendency to foster emotional bonds with inanimate objects by treating them as if they were sentient beings.

Fungi, however, come with no prior sense of identification associated conceptually to them. In spite of their increasing popularity as cultural tropes, there is still a limited lexicon which ascribes them as cute, playful, endearing or intelligent. This weirdness, combined with their intensely successful ability to conquer 'an astonishingly wide range of habitats, fulfilling an important role in diverse ecosystems' (Webster & Weber, 2007, p. 2) due to their sheer variety, gives them a massive amount of scope in imagining their potential contribution. This lack of behavioural familiarity focuses the attention of their purpose, characteristics and form factor. As an alternate metaphor, we can disrupt, if not escape, the human desire to design and recognize as users certain forms and agents of intelligence as familiar or safe for being modelled on comfortable qualities. With particular interest in the home and smart objects in this context, we focus here on what the metaphor can offer in terms of AI in everyday life. The biological processes of replication, decay, ingestion and excretion – and ultimately survival – have some interesting points of provocation as starting points to imagine future digital living.

As non-ambulatory colonies, concepts such as movement become reconfigured around how their stationary qualities can be used as displays, such as in the *Circularis* example. For the case of Doughnut worm, the distinctive visual markers are a symptom of infection, the visuality of this signalling on the body becoming a useful tool for repurposing. Fungi, like other organisms, favour optimal conditions to become established and thrive. Within the fictitious *Circularis conspicio*, the visuality of the symptom has been considered as a property. For example, diabetics' blood levels may fluctuate to abnormal and dangerous levels. Continuous glucose monitoring (CGM) devices or embedded insulin pumps monitor blood sugar levels from sensors either worn or inserted under the skin. CGM devices in particular feed levels to a display carried by the user (NHS, 2019). Using the ringworm form, this is reimagined into a functional bodily display. Responding to the host as the growth medium, the rings appear with varying prominence on the skin to provide a worn marker of hypo- or hyperglycaemia as AI monitors the levels, providing analysis on an individual basis. This plays with the concept of the host body as usually unwilling, and fungal infection as unwelcome.

Alternatively, we can question how observed traits, such as the Inky Cap mushroom which breaks down into black sludge in a form of autodigestion, can be rethought into useful, deliberately generative ways. This speculation resulted in *Coprinus notitia deliquesce* (Shaggy data cap) in the spotting guide. In this instance, we have explored how a fungal AI might collect movement data and use this to influence the actions of a networked smart thermostat (which learns to adapt to the data passed to it). However, in the absence of data for a specified period of time (which might be set to a number of months, for example), the fungus will begin autodigestion as a privacy-preserving measure. We have used the concept of not only generation but also the decay and lifecycle of fungus as prompts.

AI, as a set of algorithms, is formless. Aside from the hardware to run systems and the sensors required to monitor an environment, there is no required form, shape or aesthetic which is inherently necessary. The high degree of environmental specificity of fungi, combined with their use of existing substrate material, means that they come in a myriad of types. This flexibility can be borrowed to avoid the tropes of 'housing' artificial intelligence in a casing and placing the new object into a home, and instead privileging the integration of AI into existing dwellings, with the abode itself as the substrate medium.

In our imagined *Mycelium construe*, we extended this concept to explore if the nutrients provided by breaking down the substrate could in fact be data in place of biological molecules, and how this could feed the growth of fungi as colonies. *Mycelium construe* presents this by considering the properties of data as embodied. Data generation from home monitoring is continuous. For an AI which analyses and cross-references as it learns from listening to conversations within the home, this creates a wealth of new data as it processes individual small parts within larger work packages. This output is based on fungi which are individually tiny but occur in collectives. These create expanding, irregular organic forms composed of small component fungi. This presents further questions surrounding AI and if it can be reconsidered as a colony or collective which serves a function, as opposed to an overall singular intelligent system. Equally, with *Mycelium construe* there is a proposed traversal of the physical and digital with a smart object using excess digital nutrients to form physical artefacts which might then be of use in the home. This idea of the fungiform offering a functionally useful by-product that can be harvested by people is further seen in both *Lacrymaria digitalis* and *Cultura pomum* finding further novel uses of spare data.

Some fungiforms have somewhat sinister overtones. With *Ophiocordyceps roombatis* (zombie cleaner fungus) we are taking direct inspiration from a very curious function of some fungi that specifically act by taking over hosts' bodies. There are classes of fungi, predominantly in the rainforest, which infect insects, such as leaf-cutting ants, and then take control of their brains. Through complex biochemistry they force the ants to come down from the tree canopy to the forest floor – where climatic conditions are more appropriate for fungal growth, before compelling the ant (through manipulation of neurotransmitters) to identify an appropriate plant stem to latch onto in a permanent bite. The sap from the plant stem, pumping through the insect, provides nutrients for the fungus to grow further. The ant dies, and the fruiting body of the fungus erupts from the insect's brain before releasing its spores in to the atmosphere to propagate itself. With this in mind we have drawn inspiration from the way in which a fungal AI might agentically take control of autonomous networked devices in the home that have mobility, such as smart vacuum cleaners. Such an AI might be cultivated, however, to do this for prosocial purposes, for example, by driving around the house to identify new networked objects (other smart devices) that are visibly in the house (as opposed to being close but in an adjacent neighbour's house). This activity could be repurposed so as to automatically configure and add new domestic devices to the home network, with limited human intervention. The means of propagation we highlighted above was also directly modelled on the principles of the zombie ant fungus, suggesting how data might be exchanged between hardware, at the end of a service life, as a spore cloud of data points for exchange.

In an additional example, *Mycorrizhae salus*, we leveraged the notion of the mycorrizhae network, further highlighting that property of fungus that connects different intelligent organisms together in mutually beneficial exchanges and in ways that extend beyond what we might normally identify as the boundaries of a system. In this case, by using light-based communication to alert and signal other systems/networks about security breaches, without requiring risky network access to those systems.

Whilst we take quite literal inspiration in these examples, our work highlights that fungi are complex. What we have mirrored here is that fungi exist in rich and diverse ecosystems, with biodiverse surroundings. These, per the symbiotic relationships with plants in the mycelium networks, interact with and respond to both each other and changes in the environment.

Taking inspiration from fungi, we have deployed individual characteristics which are of particular note or interest. The sheer diversity of fungi means that our imagined examples could be expanded, altered or adapted into any number of permutations or alternatives. What the alternative metaphor provides is a set of criteria which are not closely aligned to the familiar. The peculiar and otherly nature of fungi provide new considerations for how AI may be useful. This provocation provides a different starting point and thus perhaps arrives at a different end point, to more common paths which are shaped by zoomorphic or anthropomorphic underpinnings.

In conclusion – extending the play of possibilities of AI

This chapter aims to provoke thought and dialogue around the design of everyday AI both now and in the future. Having introduced our field notes of fungal AI, and then having drawn out a little further some of the bio-'insporations' (*sic*) that they have been based upon, we turn to briefly reflect on what values such an exploration may have held for us. How have we extended the play of possibilities for the design of domestic smart objects through the use of a fungal AI metaphor? We list, below, eight characteristics of fungal AI, which we might not otherwise have derived without this proto-exploration. This is not to say that these ideas are so startlingly original that we might not have achieved them otherwise, or that we are being exhaustive herein with our exploration of the possibilities of fungal AI. This is merely to suggest eight values that we might wish to further explore in the design of smart domestic objects, which our exploration of fungal AI has specifically helped to frame.

1. *Learnt accountability of physical activity*: A common mistake with domestic AI is to assume that we must immediately know everything that it will do, and know how to read it. A shift in temporality and a consideration of fungal timescales would suggest new ways of thinking about how we design for learning behaviours over time and allow for adaptation of both user and AI. This is of course essential, when thinking through the temporalities of designing for the lived experience of the home. The timescales of domestic occupation and the objects and appliances with which we live can be extensive.
2. *Permeating boundaries of the home and the exterior*: Commonly we think of domestic AI as being situated in the home – possibly extending to connections with the internet through services such as grocery shopping

– but why not think about what domestic AI might do as a commonwealth resource for passers-by?

3. *Data residua*: Anonymized footprints of occupant activity data within the built environment that support the growth of new behaviour-responsive algorithms. Thinking through the fungal life cycle creates new concepts for how AI might deal responsibly with data in the home when we remember that there is a value to data-in-place even if the occupants move on.
4. *Prosocial control of other smart objects*: Again we commonly think of domestic AI embedded and embodied within a singular object or device, but there may be useful reasons why we would want the AI to be able to embody and interact with other devices or a sequence of devices.
5. *Alternative digital-physical networks between 'smart' homes*: The boundaries of the smart home will bleed in to one another. The overlapping nature of networks poses new kinds of security risks which might require new kinds of security solutions. Mycorrhizal fungal networks might offer inspiration for new kinds of networking for security.
6. *On-the-body AI*: Frequently we think of smart domestic objects as AI that is embedded within the home and externalized from the user – embodied in smart devices and objects – but there are good reasons why we might think alternatively about the embedding of AI into people, within symbiotic relationships of mutual benefit. This moves us beyond the cyborg notions of 'added functionality' through implant technologies to something more sophisticated through the introduction of concepts of mutuality.
7. *Comparative exchange*: Providing a harvestable value for data exchange offers new economic ways of thinking about the kinds of service relationships we might have with smart objects in the home.
8. *Novel physicalization of the digital*: Thinking through ways in which abstract data might be given form and therefore utility may offer entirely novel ways of thinking through the smart reuse of extraneous data spillage in the home.

The range of concepts from the fungal world that we have explored – growth, adaptation to an environment, symbiosis, access to resources and optimal conditions to thrive – are those which may have interesting outcomes when considered in terms of AI. Here they have resulted in a series of individual AI fungal forms which explored novel concepts for the home and its data. Within the everyday, the nature of embodiment which houses a form of created intelligence can benefit from consideration in radical and unusual forms. Thus, the forms here not only serve as individual speculations but also highlight that moving away from anthropomorphic and zoomorphic forms can generate new perspectives on functionality, application, human-AI relationships and, more broadly, the form factors of the kinds of domestic smart objects that we design. Hopefully, this chapter has provided some food for thought on the importance of carefully selecting the metaphors from which we draw and anchor our designs, and has shown how taking an alternative approach can push design in new and vital directions.

Bibliography

Anderson, R. (1994). Representations and requirements: The value of ethnography in system design. *Human–Computer Interaction*, *9*(2), 151–182.

Babikova, Z., Gilbert, L., Bruce, T. J., Birkett, M., Caulfield, J. C., Woodcock, C., *et al.* (2013). Underground signals carried through common mycelial networks warn neighbouring plants of aphid attack. *Ecology Letters*, *16*(7), 835–843.

Boa, E. (2004) *Wild edible fungi: A global overview of their use and importance to people.* Food and Agriculture Organization of the United Nations. Retrieved 22 February 2021 from http://www.fao.org/3/y5489e/y5489e00.htm.

Boston Dynamics (2004). https://www.bostondynamics.com/legacy.

Breazeal, C. L. (2002). *Designing sociable robots.* Cambridge, MA: MIT Press.

Dereshev, D. (2018). *Smart wonder: Cute, helpful, secure domestic social robots.* Doctoral thesis, Northumbria University.

Dix, N. J., & Webster,J. (1995). Aquatic fungi. In *Fungal Ecology*. Dordrecht: Springer. https://doi.org/10.1007/978-94-011-0693-1_9.

Fernaeus, Y., Håkansson, M., Jacobsson, M., & Ljungblad. S. (2010). How do you play with a robotic toy animal?: A long-term study of Pleo. In *Proceedings of the 9th International Conference on Interaction Design and Children (IDC '10)* (pp. 39–48). New York: ACM.

Forlizzi, J., & DiSalvo, C. (2006). Service robots in the domestic environment: A study of the roomba vacuum in the home. In *Proceedings of the 1st ACM SIGCHI/SIGART conference on Human-robot interaction (HRI '06)* (pp. 258–265). New York: ACM.

Gadd, G., Watkinson, S., & Dyer, P. (Eds). (2007). *Fungi in the environment (British Mycological Society Symposia).* Cambridge: Cambridge University Press.

Griffin, A. (2018, 8 May). Google Duplex: Company reveals 'terrifying' artificially intelligent bot that calls people up and pretends to be human. *Independent.* Retrieved 24 November 2019 from https://www.independent.co.uk/life-style/gadgets-and-tech/news/google-duplex-ai-artificial-intelligence-phone-call-robot-assistant-latest-update-a8342546.html.

Hawksworth, D. L. (2001). The magnitude of fungal diversity: The 1.5 million species estimate revisited. *Mycological Research*, *105*(12), 1422–1432.

Jacob, R., Girouard, A., Hirshfield, L. M., Horn, M., Shaer, O., Treacy Solovey, E., *et al.* (2008). Reality-based interaction: A framework for post-WIMP interfaces. In *Proceedings of the SIGCHI Conference on Human Factors in Computing Systems (CHI '08)* (pp. 201–210). New York: ACM.

Milenge Kamalebo, H., Malale, H. N. S. W., Ndabaga, C. M., Degreef, J., & De Kesel, A. (2018). Uses and importance of wild fungi: Traditional knowledge from the Tshopo province in the Democratic Republic of the Congo. *Journal of Ethnobiology and Ethnomedicine*, *14*(1), 13.

Mori, M., MacDorman, K. F., & Kageki, N. (2012). The uncanny valley [from the field]. *IEEE Robotics & Automation Magazine*, *19*(2), 98–100.

NHS (2019). Continuous glucose monitoring (CGMs). https://www.nhs.uk/conditions/type-1-diabetes/continuous-glucose-monitoring-cgms//.

Norman, D. (2013). *The design of everyday things (revised and expanded edition)*. Cambridge, MA: MIT Press.

Nowacka, D., Hammerla, N., Elsden, C., Plötz, T., & Kirk, D. (2015). Diri – the actuated helium balloon: A study of autonomous behaviour in interfaces. In *Proceedings of the 2015 ACM International Joint Conference on Pervasive and Ubiquitous Computing (UbiComp '15)* (pp. 349–360). New York: ACM.

Rowland, C., Goodman, E., Charlier, M., Light, A., & Lui, A. (2015). *Designing connected products: UX for the consumer internet of things*. Sebastopol, CA: O'Reilly Media Inc.

Sung, J. Y., Grinter, R., & Christensen, H. I. (2009). 'Pimp my Roomba': Designing for personalization. In *Proceedings of the SIGCHI Conference on Human Factors in Computing Systems (CHI '09)* (pp. 193–196). New York: ACM.

Vaussard, F., Fink, J., Bauwens, V., Rétornaz, P., Hamel, D., Dillenbourg, P., *et al.* (2014). Lessons learned from robotic vacuum cleaners entering the home ecosystem. *Robot. Auton. Syst.*, *62*(3), 376–391.

Webster, J., & Weber, R. (2007). *Introduction to fungi (3rd ed)*. Cambridge: Cambridge University Press.

2 DRAMATURGY FOR DEVICES: THEATRE AS PERSPECTIVE ON THE DESIGN OF SMART OBJECTS

Maaike Bleeker and Marco C. Rozendaal

In this chapter we reflect on how insights and expertise from the theatre can inform the conceptualization and design of smart objects in everyday life. In the field of human-computer interaction (HCI), theatre has a history of being referred to as a generative metaphor in the design of user interfaces of computing systems. In her pioneering work, Brenda Laurel (1993) proposes theatre as model for interface design and for navigating the virtual. She developed her perspective on 'computers as theatre' in the midst of the multimedia revolution and in relation to the virtual worlds existing within the computational spaces opened up by computer interfaces. Laurel shows how insights from the theatre, in particular Aristotle's poetics, are most useful for what she describes as 'a dramatic theory of human-computer interaction' (xvii). Building on the tradition set by Laurel, we too propose theatre as a perspective on design, albeit not of computer interfaces but of smart objects and their modes of performing. Unlike the virtual, other worlds opened up by computer interfaces, smart objects exist and operate within the real material world of users. In this context, we will show not Aristotle's theory of dramatic narrative, but how dramaturgical concepts and insights regarding staging situations in the here and now can provide designers with conceptual tools to understand and design the interaction between humans and smart objects embedded in shared environments.

The term 'dramaturgy' is used to refer to the totality of all aspects that are part of how theatre performances are constructed, the relationships between these elements and how these relationships unfold in time and space. This may involve storytelling (as in dramatic plays), but not necessary so. Performances can also be organized according to other logics and other compositional principles such as that of montage, visual composition, choreography or gamelike structures. Performances can be constructed to take the audience along in experiences and associations by means of compositions of materials that do not tell a story or represent another world but set up a situation in the here and now. Doing dramaturgy in the context of the theatre involves paying attention to how performances do what they do as a result of how they are constructed. Dramaturgy thus understood is not itself an approach to designing performances, but rather consists of a set of tools, terms and insights to think through the logic of (real or fictional) situations and how they afford interactions and interpretations, suggest interpretations and trigger actions and associations.

In traditional Western theatre – that is, the kind of theatre that is based on the ideas of Aristotle as used by Laurel – the various elements of theatrical staging and the ways in which they are brought together are used for the representation of fictional worlds. Like computer interfaces, the means of the theatre here serve first and foremost to provide access to 'virtual' other worlds. Ever since the early twentieth century, however, avant-garde theatre makers have developed new strategies of creating theatre that foreground the here and now of the theatrical event, its materiality and embodiment, and its mode of addressing the audience in shared time and space. In such theatre, attention shifts away from narrative and representation of other worlds (central dramaturgical principles of Aristotelian theatre as used by Laurel) and towards the composition of human and non-human performers, the things, sounds, texts and movements that together make up the theatrical performance. In the following, we will show how insights in these aspects of theatrical performance may support an ecological approach to the design of smart objects that starts from the relationships defining the situation in which the smart object is to operate in and from how the object negotiates these relationships, building on them, or intervening in them, and transforming them. We will also discuss how this approach invites a reconsideration of what constitutes the 'objecthood' as well as the 'smartness' of smart objects in relational terms. We will do so using the project *Mokkop* as our design case. Although dramaturgy was not part of the design process of Mokkop (see Vermeeren, van Beusekom, Rozendaal & Giaccardi, 2014), we will use Mokkop as our example of how dramaturgical concepts and insights such as 'mise-en-scène', 'performativity', 'presence' and 'address' may support further understanding of what happens in the design process and may contribute to further developing ecological approaches to design.

FIGURE 2.1 Impression of *Mokkop*.

Design case: *Mokkop*

Mokkop was designed by Josje van Beusekom in the context of her master graduation project at Delft University of Technology in collaboration with the Princess Máxima Centre for Pediatric Oncology. The aim of the project was to develop a product to support parents of hospitalized children to take time for themselves (Figure 2.1). Children diagnosed with cancer often have to stay in the hospital for extended periods of time. It was observed that the caregivers staying with them, focusing all their attention on the children and being absorbed in care and worries about them, tend to become lonely and isolated. *Mokkop* was developed as an intervention that aimed to prevent this from happening, or at least make it less extreme. To this end, van Beusekom created a series of coffee cups that have the capacity to glow and show intricate patterns of light at various moments during the day. Parents and other caregivers accompanying small children who are hospitalized are invited to select a cup of their choosing. The cup looks quite ordinary for most of the time, apart from five times a day when it starts to glow. Caregivers are invited to take this as a suggestion that it might be time for a cup of coffee, or a small break. At the same time that the cup of one caregiver starts to glow, the cup of another caregiver in a nearby hospital room does so too. The cups thus make an intervention in the given situation that sets the stage for taking a break and for a meeting at the coffee machine. Caregivers are not ordered or forced to actually go to the coffee machine or engage in a chat. Rather, this possibility is implicated in the design as a potential for action and experience, an invitation they can choose to respond to.

An ecological approach to design

Mokkop is an interesting example of what can be called an ecological approach to design: an approach that start not from designing an autonomous entity to be put into an environment, but rather from how the object to be designed might tap into not yet realized potential immanent to the environment and actualize it. Such an ecological approach to design is not necessarily about consideration for the environmental impacts of the product (as proposed by Van der Ryn & Cowan, 1995). Rather, it follows Félix Guattari's (2000) extension of ecology to encompass social relations and human subjectivity as well as the material environment, and looks for ways to actualize potential immanent within this ecology. Peter Trummer (2008), developing Guattari's (2000) elaborations on ecological thinking as an approach to engineering, describes this immanent potential as the virtual dimension of the environment. Virtual, thus understood, is not something fictional. In line with Gilles Deleuze's understanding of the virtual, both the actual and the virtual are fully real. The former has concrete existence while the latter does not, yet (Buchanan, 2010).

The challenge of ecological design then is to recognize possibilities that are virtually present within what is actually there, how these may provide solutions to design questions and how these virtual possibilities can be actualized by means of design interventions in the environment. This is how we may describe what happened in the design process of *Mokkop*. The object of design here is not merely a thing (the cup) that is then put into an environment, but the intervention made by the cup that mediates in actualizing the potential for taking a break and for having a chat with a fellow caregiver at the coffee machine. The design of *Mokkop* involves quite literally an intervention (the lighting up performed by the cup) that invites to actualize a not-yet-realized potential given in the set-up of the environment. Such actualization is truly ecological in that what is designed and the environment are part of the same becoming.

Insights from theatre may contribute to developing awareness of such potentiality and to how it can be put to use for design purposes. Making theatre is all about setting the stage for the emergence of what is not yet there, about recognizing the potential of relationships between people and things within situations, and the potential of well-chosen actions to intervene in situations in ways that set them into motion. Furthermore, theatre is all about doing so in relation to the expectations and assumptions of human users, called spectators. Theatre is constructed with spectators in mind and in order to invite certain ways of understanding rather than others, to trigger certain emotions and associations rather than others, and (in certain types of theatre) even to make spectators do certain things rather than others (for example, in participatory theatre, theatre of experience, or other types of theatre that require actual activity of the audience).

Making theatre requires understanding of how to play into and play with expectations, conventions and culturally and historically specific ways of looking, doing and understanding. Finally, theatre provides a relevant model for designing smart objects in how it redirects attention with regard to what the object of design is. A theatre performance consists of a great number of different elements (including things, humans, texts, movements, sounds, etc.), and its objecthood is to a large extent given in how the theatrical apparatus sets up relationships between these elements and with an audience. Similarly, we might argue that what is being designed when designing a smart object is not merely an autonomous thing, for example, a cup, but how this cup actualizes relationships within an environment. Designing such objecthood requires precise insight in what in dramaturgical terms is called 'mise-en-scène'.

Mise-en-scène

Mise-en-scène describes in a broad sense the arrangement of 'all of the resources of stage performance: décor, lighting, music and acting'. In a narrower sense, 'the term "mise-en-scène" refers to the activity that consists in arranging, in a particular time and space, the various elements required for the stage performance of a dramatic work' (Pavis, 1998, p. 363). Mise-en-scène can thus be used as an analytical term that draws attention to the specificities of this arrangement in an already existing theatre performance. It can also describe the practice of creating such arrangements. In both cases, mise-en-scène draws attention to the composition of the arrangement in time and space and how this arrangement affords the unfolding of action. Mise-en-scène provides a conceptual tool for designing smart objects that directs attention to the situation in which the smart object is to operate as a spatio-temporal arrangement of humans and things, as well as to the identified relationships between the arrangement of the situation and the actions and experience of the people within this physical and social context, their interdependencies (human, object and environment) concerning the state-of the world as it is as well as how it could be like.

The design process of *Mokkop* began with an investigation of what we might call the mise-en-scène of the situation of the caregivers in the hospital and how this mise-en-scène sets the stage for their actions. In the design process this involved conducting user research that helped designer van Beusekom to understand hospitalization from the perspective of the family, concerning their needs, experiences and problems they face. A careful analysis of the hospital architecture provided van Beusekom with an understanding of the physical layout of the patient, room, waiting areas and hallways. Observations being done in the hospital provided additional information about how the setting relates to the activities that take place there, and interviews provided information about the feelings they trigger. Van Beusekom developed her design vision by identifying the barriers such as the current situation of the child, having a reason to carve me-time, finding the right moment as well as what might help caregivers to take a moment to relax. These insights in the current situation and the identified opportunity for an intervention that might actualize a still unrealized potential of this situation set the stage for *Mokkop* to crystallize as a design concept: a cup as an actor capable of bringing about the desired change in the given situation. This capacity of bringing about a new situation is what we may call the performativity of the cup.

Performativity

Performativity entails understanding how what kind of behaviour of what kind of object may bring about the desired change within a given mise-en-scène. This understanding of performativity is based on speech-act theory as introduced by John Austin (1975) and John Searle (1969) and further elaborated by Judith Butler (2007) and, more recently, by (among others) Jon McKenzie (2001) and Karen Barad (2007), drawing attention to the performativity of technology (technoperformance) and presenting a post-humanist perspective on performativity. These theories help to understand that saying things and doing things have the power to 'bring about' things within the situation in which they are performed. The canonical example of speech-act theory is that of the wedding vow transforming two unwedded people into a married couple. This power of words and actions to bring about identity and situations, as well as intervene in them and transform them, is well known to theatre makers too. With only a few words whole worlds can be evoked on stage. Puppet- and object theatre makers have shown that well-chosen movements can produce a sense of character and 'aliveness' in almost any kind of object. Playwrights like Chekhov have demonstrated how simply the entrance of a character at a well-chosen moment can be an intervention that completely changes the entire situation on stage.

Performativity is not a matter of what an object (like e.g. the cup in *Mokkop*) does per se (its performance) but describes what this doing brings about within the given situation. This requires understanding this object as an agent ('actant', Latour) within a network (apparatus) of relationships of multiple human and non-human agents that mutually influence each other, and in which the object can act as a mediator in complex physical and social settings. Following Barad, we might say that it is within the context of a given apparatus (i.e. within a particular network of relationships between humans and things) that an object (like the cup in *Mokkop*) gains the agency to intervene and bring about a change. Designing a smart object, therefore, requires recognizing what kind of behaviour of what kind of object has the potential of mediating in bringing about the desired change. It also involves recognizing that what may appear as the 'character' and the intentions of a smart object are the effect of what it does and how, and how this can be interpreted within the given situation.

Mokkop demonstrates that the performativity of the smart object is not a matter of an object successfully expressing the intention of bringing two people together at the coffee machine for a chat. Rather, key to its design was figuring out what kind of performance of what kind of object would be able to generate the desired action of the caregivers. Furthermore, the agency of the cup (its capacity to bring about this change) is inseparable from the situation. The same behaviour in a different situation would not necessarily have this potential. The performativity of *Mokkop* is the result of a combination of the choice for a cup, the design of the cup, its specific way of performing and the situation in which this happens (including the availability of the coffee machine, the material arrangement of the refreshment room and the presence of more than one caregiver). That is, it involved all kinds of things that are crucial for the intervention to have the desired effect but are not directly part of the design of the cup as object in itself. Achieving the desired behaviour of the caregivers also involved working out the 'time table' of different sets of cups glowing at different moments, the rhythm of five times a day and how they should make themselves present. For although the cups are there all the time, their capacity to successfully intervene in the situation and bring about the desired effect requires them to draw attention to their presence five times a day.

Presence

In his *Dictionary of the Theatre*, Patrice Pavis observes that 'to have presence' in theatre jargon means knowing how to 'captivate the audience' (1998, p. 285). Presence is an ambiguous concept in how it is associated both with something some people know how to do (knowing how to captivate the audience) and with a quality that one simply has or has not. It is certainly the case that some actors manage to captivate their audiences better than others, while this capacity also appears to be context dependent. Actors endowed with an impressive stage presence do not necessarily have a similar strong presence in daily life. This seems to suggest that their presence is situated and related to the context of the stage. Yet, although stage presence can be enhanced by theatrical means like, for example, light (putting someone or something in the spotlight), composition (like in ballet, where the composition of the corps de ballet directs the attention of the audience towards the soloists), costume or the performance of co-performers (like the stooge in a comic duo), some people and some things seem to be better capable of captivating the audience than others.

Different modes of presence and the effect of ways of increasing presence are also an important part of the design of *Mokkop*. Central to *Mokkop*'s mode of operating is the precisely timed and organized becoming present of the cups five times a day, and how this transformation of their presence presents an invitation to the caregivers. The cup is there all the time, but crucial to its modes of operating is that it makes itself present at specific moments and how such presencing draws attention to its existence. Using light rather than sound or movement as a means of increasing the presence of the cup affords a non-invasive way of drawing attention. Sound could easily disturb the precarious situation in the hospital room, wake up the patient or be experienced as annoying when happening at a less appropriate moment. The glowing of the cup can more easily be ignored if happening at a moment the caregiver does not want to respond. If desired, it can do its job without attracting the attention of the patient, especially when placed outside their field of vision. The manner in which the cup makes itself present also implies ways of engaging people in how it invites them to relate to it, understand it and do things with it. This is what is called 'address'.

Address

Insight into how behaviour and looks set the stage for possible responses and thus for modes of interacting with fellow actors as well as with the audience is an important skill for actors. Acquiring such skills involves developing an understanding of how the way one addresses fellow actors or the audience invites, triggers and makes possible certain ways of responding, while foreclosing others. Modes of address affect how the one being addressed is invited to understand what is shown and done, is invited to sympathize or not and so forth. Furthermore, staging involves understanding how not only the actors but also all that is shown and done on stage does things to spectators: how this address evokes a sense of self in the situation and invites ways of responding and understanding. The theatre provides a model for how designing behaviour intended to achieve a particular effect requires taking into account the expectations, assumptions, desires and so forth of spectators or users.

In the case of smart objects, more than is the case with most types of theatre, address is instrumental in making people actually do things. An important question in the process of designing *Mokkop*, for example, was working out what the cup should look, feel and behave like in order for the caregivers to feel invited to pick it up and take a break for a coffee and a chat. This involved working out how the address presented by the cup implies their potential responses and how the design invites or affords ways of responding. It also involved taking into consideration how caregivers in the hospital may feel addressed by the shape, colours and ways of lighting up of the cups, how they will 'read' ways of being addressed and how they will feel invited by them to take action. These considerations informed the choice of material of the cups and how they affect the feel of the cups, what this feel brings about and how this may affect and inform modes of social interaction.

The choice for porcelain rather than plastic was informed by how the combination of softness and strength presents an appeal to tactility and invites picking up the cup and holding it in one's hands. The choice for the specific kinds of patterns of light (abstract rather than figurative) is meant to avoid the address to respond to all too specific tastes and thus speak to some while not to others. The glowing up of patterns of light evokes associations with warmth, thus pointing forward to the possibility of the warmth and comfort of a cup of coffee or tea. Furthermore, the address presented by the lighting up of the cup is quite different from, for example, the address presented by a sound. A sound signal might more easily be associated with a command, while light easily triggers associations with something pleasurable and comforting, and the soft glowing of the cup appeared to have the effect of drawing people towards the cup.

The objecthood and smartness of smart objects: An ecological approach

Whereas Laurel's classical Aristotelian theatre-inspired approach to design invites designers to approach the creation of computer interfaces as the design of virtual, other worlds that we navigate and experience as fiction, our dramaturgy for devices – based on insights from contemporary theatre – supports an ecological approach to the design of smart object that starts from real-world interaction with people in shared environments. We have shown how this approach invites a reconsideration of the 'objecthood' of smart objects in how it shifts attention from design being a matter of giving shape to individual things to design being about relationships and about interventions with the potential to bring about changes in environments and in the behaviour of people. The objecthood of the object of design includes the relationships with the environment and with people. The example of *Mokkop* can serve as an example of an ecological approach to design in how the object of design is not merely the looks and construction of the cups but what these cups are capable of bringing about within the given environment.

Similarly, we argue, an ecological approach to the design of smart objects requires a reconsideration of the smartness of smart objects. From this perspective, the smartness of smart objects is nuanced to be not only a technical computational property of the object but also a relational quality of the object that manifests itself in interaction. Here, too, *Mokkop* can serve as an example: the smartness of *Mokkop* is a matter of the fittingness of the behaviour of the object within a specific setting and about how, within this setting, this behaviour is capable of bringing about desired effects. This understanding of smartness is at odds with the idea of the smartness of smart objects being a matter of some kind of artificial brain being implanted in autonomous objects, like Cartesian minds in machinic bodies. Even though the design of *Mokkop* does include (rather basic) electronics implanted in the cups controlling their glowing behaviour, the smartness of the object and the capacity of the design to bring about the desired effect is not a matter of how smart this electronic 'brain' is but of how the behaviour of the cups is designed to afford potentialities of the environment to be actualized. That is, smartness is not a matter of a computational system acting like a mind to an autonomous body-machine but of how the doing of the object is designed to intervene in the environment and to bring about meaningful actions of human users.

We have shown how our dramaturgy of devices supports the development of the objecthood and smartness of smart objects, and how dramaturgical insights regarding mise-en-scène, performativity, presence and address offer conceptual tools to support and further develop various aspects of ecological approaches to design. In the following, we will discuss how a dramaturgy of devices relates to current approaches and issues in the field of design, and how it may contribute to an interaction design research agenda.

Dramaturgy and interaction design

A dramaturgical approach supports a distanced yet empathic relating to a particular context and opens up the designer's eye to this situation as meaningful and complex, and as something that is enacted. As such, it contributes to design methods that emphasize human experience in relation to the social contexts and practices in which they are situated (Crabtree, Rouncefield & Tolmie, 2012; Kuutti & Bannon, 2014) and how objects situated in these contexts mediate human activity (Kaptelinin & Nardi, 2006) and ways of 'being in the world' (Verbeek, 2005). Participatory design has a rich tradition in how to involve people to help uncover the potential of what a particular situation or context could be like (Sanders & Stappers, 2012; Vines, Clarke, Wright, McCarthy & Olivier, 2013) and how this may involve consulting 'things' as well (Giaccardi, Cila, Speed & Caldwell, 2016). Contemporary theatre too has a lot to offer with regard to such collaborative creation processes involving the participation of professionals as well as non-professionals in making theatre on stage and in real-world settings. A dramaturgical approach further resonates with in-situ prototyping (Chamberlain, Crabtree, Rodden, Jones & Rogers, 2012) as being a continuous dialogue between the to-be-designed object and people in particular contexts of use, which allows its performativity to be explored and orchestrated.

Smart objects are considered to have particular kinds of agencies that scholars describe as originating from 'quasi-subjects': objects that delegate agency inscribed by others but that seemingly originate from the object itself (Latour, 1993; Bødker & Andersen, 2005). Designing interactions with such objects requires careful thought about who or what has the initiative and who or what controls how the interaction unfolds. Notions such as 'negotiation' (Frauenberger, this book), 'collaboration' (Rozendaal, Boon & Kaptelinin, 2019) and 'co-performance' (Kuijer & Giaccardi, 2018) are concepts that address these particular agencies and acknowledge how outcomes of the interaction are co-produced. Here, a sense of humbleness is warranted. As designers are not able to dictate how smart objects will exactly enable change but can only influence how the object invites to action, designers should think about how smart objects speak to human creativity and how they allow for improvisation and appropriation (D'Olivo, van Bindsbergen, Huisman, Grootenhuis & Rozendaal, 2020). Here, dramaturgical insights in (already mentioned above) 'address' and 'performativity', as well as a rich body of knowledge regarding the staging of interactions and various types of improvisation could provide useful additions to the designer's toolkit and support an understanding of interaction as an emergent phenomenon in which humans and objects participate.

An important aspect of organizing interactions between humans and smart objects is timing. Interaction episodes between humans and objects are of a certain duration and can be repetitive over time. To become embedded in particular contexts, objects need to act in synchronization with human behaviour and adapt to ongoing activities and routines. This indicates how temporal aspects of smart objects are both linear and cyclic (Engeström, 1999). For instance, linear temporal aspects may relate to how an object's intent and intelligence is expressed in interaction through its form and behaviour (Vallgårda, 2014) while cyclic temporal aspects relate to how an object can establish a presence over time by interacting and being present in particular moments. This further alludes to thinking about what meaning objects have 'in-between' interaction episodes (Odom *et al.*, 2014). Time, timing and temporality are also important to the construction of theatrical performances and how they engage audiences. Expertise from the theatre with regard to ways of structuring time, marking of time, rhythm, expectation management and 'attunement' would make most useful contributions to a smart objects designer's toolkit.

How to design the character of smart objects? Lars-Erik Janlert and Erik Stolterman (1997) define the character of artefacts as the unity of an object's multiple characteristics, which involves the sustained impressions of aspects of the object's function, appearance and manner of behaving, aggregated over time in a complete and coherent way. This relates very well to dramaturgical insights in character as an emergent phenomenon. Like smart objects, characters on stage do not have a pre-existing inner identity that expresses itself. What appears as character is brought about by what they look like and what they do and how others respond to them. In the theatre, characters are often performed by humans, but not always. Objects can be performers too. The character of Tinkerbell in Peter Pan, for example, is traditionally performed by a light. Object theatre makers have shown how all kinds of objects can be turned into partners in performance. The skills and aesthetic sensibilities required to animate objects are increasingly recognized as valuable assets in the design of smart objects and social robots (Hoffman & Ju, 2012; Bianchini, Levillain, Menicacci, Quinz & Zibetti, 2016; D'Olivo, Rozendaal & Giaccardi, 2017). Here too, it seems, a lot is to be gained from bringing together dramaturgical expertise and interaction design.

Conclusion

Brenda Laurel's *Computers as Theatre* has greatly influenced the design of human-computer interaction and contributed to the development of virtual worlds in ways that are more human, emotional and understandable. With our dramaturgy for devices we propose to take her approach in a new direction and draw attention to the potential of knowledge embodied in more contemporary and less representational types of theatre for the design of smart objects. Inspired by Laurel's pioneering work, we propose combining insights from the theatre with competencies and skills that trained interaction designers are familiar with, as a useful addition to the smart object designer's toolkit. A dramaturgy for devices supports an understanding of smart objects as entities that actively form relations within ecologies of people and things, intervene in such ecologies and bring about changes as a result of these interventions. It also supports an understanding of the intelligence of smart objects as given in their ability to establish relationships and effect transformations. Insights from the theatre have a lot to offer for the further development of such relational approaches to the objecthood and smartness of smart objects in ways that do not start from mimicking human or animal intelligence but rather from how the smart object inhabits an ecology of relationships. Being designed to both fit in and actualize unrealized potential of these ecologies, smart objects present an image of intelligence and of agency as inseparable from the environment and from the entities' potential for (inter)action within it. This is an understanding of intelligence and of agency in line with Latour and Barad's observations on how it is from its being part of actor-networks or apparatuses consisting of human and non-human elements that entities gain agency. With our dramaturgy for devices we propose the theatrical apparatus as a model to think through the inseparability of the smartness of the smart object and the ecology within which the object operates, and expertise from the theatre as a rich source of knowledge about creative engagements with ecologies and their performative inhabitants.

Bibliography

Austin, J. L. (1975). *How to do things with words.* Cambridge, MA: Harvard University Press.

Barad, K. (2007). *Meeting the universe halfway: Quantum physics and the entanglement of matter and meaning.* Durham, NC: Duke University Press.

Bianchini, S., Levillain, F., Menicacci, A., Quinz, E., & Zibetti, E. (2016). Towards behavioral objects: A twofold approach for a system of notation to design and implement behaviors in non-anthropomorphic robotic artifacts. In J-P. Laumond & N. Abe (Eds), *Dance Notations and Robot Motion* (pp. 1–24). Cham: Springer.

Bødker, S., & Andersen, P. B. (2005). Complex mediation. *Human-computer Interaction, 20*(4), 353–402.

Buchanan, I. (2010). *A dictionary of critical theory*. Oxford: Oxford University Press.

Butler, J. (2007). *Gender trouble: Feminism and the subversion of identity*. New York: Routledge.

Chamberlain, A., Crabtree, A., Rodden, T., Jones, M., & Rogers, Y. (2012, June). Research in the wild: Understanding 'in the wild' approaches to design and development. In *Proceedings of the Designing Interactive Systems Conference* (pp. 795–796). New York: ACM.

Crabtree, A., Rouncefield, M., & Tolmie, P. (2012). *Doing design ethnography*. London: Springer Science & Business Media.

Engeström, Y. (1999). Activity theory and individual and social transformation. *Perspectives on Activity Theory, 19*(38), 19–30.

Giaccardi, E., Cila, N., Speed, C., & Caldwell, M. (2016, June). Thing ethnography: Doing design research with non-humans. In *Proceedings of the 2016 ACM Conference on Designing Interactive Systems* (pp. 377–387). New York: ACM.

Guattari, F. (2000). *The three ecologies*. London: Bloomsbury.

Hoffman, G., & Ju, W. (2012). Designing robots with movement in mind. *Journal of Human Robot Interaction, 1*(1), 78–95.

Janlert, L. E., & Stolterman, E. (1997). The character of things. *Design Studies, 18*(3), 297–314.

Kaptelinin, V., & Nardi, B. A. (2006). *Acting with technology: Activity theory and interaction design*. Cambridge, MA: MIT Press.

Kuutti, K., & Bannon, L. J. (2014). The turn to practice in HCI: Towards a research agenda. In *Proceedings of the 32nd Annual ACM Conference on Human Factors in Computing Systems – CHI '14* (pp. 3543–3552). New York: ACM.

Kuijer, L., & Giaccardi, E. (2018, April). Co-performance: Conceptualizing the role of artificial agency in the design of everyday life. In *Proceedings of the 2018 CHI Conference on Human Factors in Computing Systems* (pp. 1–13). New York: ACM.

Latour, B. (1993). *We have never been modern*. Cambridge, MA: Harvard University Press.

Latour, B. (2005), *Reassembling the social: An introduction to actor-network-theory*. Oxford: Oxford University Press.

Laurel, B. (1993). *Computers as theatre*. Reading, MA: Addison-Wesley.

McKenzie, J. (2001). *Perform or else: From discipline to performance*. New York: Routledge.

Odom, W. T., Sellen, A. J., Banks, R., Kirk, D. S., Regan, T., Selby, M., *et al.* (2014, April). Designing for slowness, anticipation and re-visitation: a long term field study of the photobox. In *Proceedings of the SIGCHI Conference on Human Factors in Computing Systems* (pp. 1961–1970). New York: ACM.

D'Olivo, P., Rozendaal, M. C., & Giaccardi, E. (2017, June). AscoltaMe: Retracing the computational expressivity of a tactful object for sensitive settings. In *Proceedings of the 2017 Conference on Designing Interactive Systems* (pp. 943–955). New York: ACM.

D'Olivo, P., van Bindsbergen, K. L. A., Huisman, J., Grootenhuis, M. A., & Rozendaal, M. C. (2020). Designing tactful objects for sensitive settings: A case study on families dealing with childhood cancer. *International Journal of Design, 14*(2). Retrieved 12 March 2021 from http://www.ijdesign.org/index.php/IJDesign/article/view/3537/911.

Pavis, P. (1998). *Dictionary of the theatre: Terms, concepts, analysis*. Toronto: Toronto University Press.

Rozendaal, M. (2016). Objects with intent: A new paradigm for interaction design. *Interactions, 23*(3), 62–65.
Rozendaal, M. C., Boon, B., & Kaptelinin, V. (2019). Objects with intent: Designing everyday things as collaborative partners. *ACM Transactions on Computer-Human Interaction (TOCHI), 26*(4), 1–33.
Sanders, E. B. N., & Stappers, P. J. (2012). *Convivial design toolbox: Generative research for the front end of design*. Amsterdam: BIS.
Searle, J. (1969). *Speech acts*. Cambridge: Cambridge University Press.
Trummer, P. (2008). Engineering ecologies. *Architectural Design, 78*(2), 96–101.
Vallgårda, A. (2014). Giving form to computational things: Developing a practice of interaction design. *Personal and Ubiquitous Computing, 18*(3), 577–592.
Van der Ryn, S., & Cowan, S.(1995). *Ecological design*. Washington, DC: Island Press.
Vines, J.,Clarke, R., Wright, P., McCarthy, J., & Olivier, P. (2013, April). Configuring participation: On how we involve people in design. In *Proceedings of the SIGCHI Conference on Human Factors in Computing Systems* (pp. 429–438). New York: ACM.
Verbeek, P-P. (2005). *What things do: Philosophical reflections on technology, agency, and design*. State College: Penn State Press.
Vermeeren, A. P., van Beusekom, J., Rozendaal, M. C., & Giaccardi, E. (2014, June). Design for complex persuasive experiences: helping parents of hospitalized children take care of themselves. In *Proceedings of the 2014 Conference on Designing Interactive Systems* (pp. 335–344). New York: ACM.

3 THE TELLING OF THINGS: IMAGINING WITH, THROUGH AND ABOUT MACHINES

Tobias Revell and Kristina Andersen

> A machine is a thing. It is on the object side of things. Yet a machine is an anomalous kind of thing, an object that seems to exceed its objecthood in certain ways, through its quality of being automatic … Machines do not work for us, and so a machine is always a kind of substitute for a subject.
>
> **Connor (2017, p. 50)**

Machines are as much imagined as they are technical propositions. Several authors, not least Gilles Deleuze, have noted that 'machines are social before being technical' (Deleuze, 1988, p. 39) and as social objects machines are bound up in the social imaginaries we create for, through, with and about them.

The nexus of the imaginary, the technical and the non-human have always been complicated and ever-shifting, riddled with apparent paradoxes. James C. Scott, David Graeber and others have written extensively on the way that machines reproduce human-centric reductionism, attempting to reduce and simulate natural phenomenon to technical processes and then reinscribing these simulations on the non-machine world (Scott, 1998; Graeber, 2015). Conversely, many others show how machines create reflexive opportunities to reconsider the relationship of the human and non-human through almost transcendental machine experiences (Pohflepp, 2016; Levitt, 2018). We imagine them to be simple tools of 'innovation', testament to human skills of exploiting natural phenomenon (Singleton, 2014) while simultaneously being rhetorical partners and meaning makers (Losh, 2016; Hayles, 2019).

These complexities demand a novel perspective on humans, non-humans and machines, which we aim to explore here. However, in seeking to briefly describe some relationships with imaginary machines and how we imagine our relationships with machines and how machines shape our imaginations and how machines imagine us and how we mechanize imagination, we will eschew the never-ending project of rationalizing complexity into technical categories. Instead, we turn to the Argentinian surrealist author Jorge Luis Borges, who, in his own satire of the absurdity of formal classification wrote (or, claimed to have discovered) the Celestial Emporium of Benevolent Knowledge, a non-Western categorization system for animals:

> In its remote pages it is written that the animals are divided into: (a) belonging to the emperor, (b) embalmed, (c) tame, (d) suckling pigs, (e) sirens, (f) fabulous, (g) stray dogs, (h) included in the present classification, (i) frenzied, (j) innumerable, (k) drawn with a very fine camelhair brush, (l) et cetera, (m) having just broken the water pitcher, (n) that from a long way off look like flies. (Borges, 1952)

Our brief review takes these categories as a starting point for problematizing and poeticizing positions and differences so that we might more easily demystify or dispel the assumed prehistoric relationship of humans, machines and imagination and see them in novel ways. Each categorization may be used as shorthand for particular forms of relationship in future work. This essay, as a whole, functions as a thing to think with.

Belonging to the emperor

> Where the emancipatory and imaginative potential of machines is foreclosed to service entrenched power dynamics.

Throughout the twentieth century, the machine was conceptualized as a thing of power. The Italian Futurists wrote their manifesto inspired by the superhuman potential of the machinic man (invariably a man). Kitschy product demonstrations of kitchen gadgets and cars laud consumerist tropes: speed, power and efficiency of production and consumption. These are the driving priorities of 'innovation', while – tragically – the word 'technology', once translatable as 'knowledge of techniques', has become synonymous with the nebulous worlds of gizmology and gadgetry bringing with it a reverence of obfuscated power, secret knowledge, access and socially constructed notions of 'progress'. Things of *Belonging to the emperor* might be described as the resting state of machines and imagination in the majority of the Western-originating cannon. They are unchallenging and uncritical, and the nexus of imagination, humans and machines tend to fit comfortably together in this category.

Belonging to the emperors are usually bound up in the imagination of individuality and superhero-like qualities enabled by technology that might enable superhuman achievement or status. When describing the emotional attraction of superheros, the cartoonist Chris Ware says, 'When we are weak we wanted to be strong, and when we were very weak we wanted to be very strong' (Ware, 2001). We can maybe extend such desires to the imagination of the powerful or magical machine. We tell stories of machines that provide the protagonist (and by extension us) with unequalled powers, an unfair advantage. The dream of the unfair advantage drives the imagination of *Belonging to the emperor* machines, real and imagined.

Embalmed

> Where machines are purely imaginary. There may be physical attempts, representations or prototypes of them but their functioning is purely imaginary.

We need look no further than television or cinema to see the world of imaginary machines in high resolution. The mainstreaming of science fiction in the 1990s and the new-found popularity of speculative fiction such as 'Black Mirror' has put discussion of the imaginary machine at the forefront of pop culture discussion of future technological innovation. For example, the recursive relationship between science fiction and innovation is apparent in technology media headlines comparing new gadgets to things previously only found in films: 'SF [Science fiction] plays an important role in the shaping of desire – for change, for progress, for novelty, for a sense of wonder and of discovery' (Bassett, Steinmueller & Voss, 2013) Think here of every headline comparing a new device or software to something from Steven Spielberg's *Minority Report* or the apparent foresight of Spike Jonze's *Her* in describing the relationship we might have with Amazon's Alexa or Apple's Siri.

Embalmed things in cinema, television and pop culture often embody foreclosed social aspirations of power and control (see *Belonging to the emperor*) but can often act as warning tales of potential technologies or even suggest new social imaginaries through imaginary machines (see *Frenzieds*). Additionally, *Embalmed* can be apocryphal or charlatan in nature as machines that claim to serve some imaginary technical function but in fact do not. Here 'the production of apocryphal technologies is fueled by desires and fantasies that can never be realized' such as mind reading or control, prediction or divination (Enns, 2019, p. 1). The lineage of charlatan or apocryphal technologies is long and storied, from the claimed healing properties of the 'animal magnetism' of Franz Anton Mesmer to bogus radiation detectors sold after the Fukushima disaster. The continued resonance of fantastic imaginary technologies is clearly a notable and relevant branch of the machine imaginary.

More pressingly, it is arguable that the imaginary function of machines is as important as their technical function. So, while the *Embalmed* might often fail to live up to their technical claims, they occupy prominent positions in individual, scientific and social imagination, making them charismatic figures in the development of new machines and imaginaries.

Tame

> Where machines are imagined to be nothing more than tools, devoid of any inherent politics or social entanglement in their construction or use.

Not much needs to be said on this category as nothing can be categorized here.

Suckling pigs

> Where the non-machine world is read as, reduced and rendered as machinic components or processes by humans.

Machines are 'constructed things' and as such are reducible to their parts. This reductionism has a tendency to pervade our worldview as it makes so much sense when dealing with the technological framework of daily social activity. Stephen Connor suggests that all human activities are 'a matter of techne, technique, or technesis' and so are 'imagined mechanically' (Connor, 2017, p.188). In other words, because all human activities are inherently technical, humans are incapable of imagining the world in a non-machinic way without classifying it, trying to reduce it to components and parts.

As Connor suggests, we can only imagine the world mechanically. Think, for example, of the reduction of the 'natural world' to a human-imposed order: the naming of species and their classification, the reduction of ecologies to machine-like systems of trees and matrices. This tendency of projecting a mechanical 'truth' onto nature through understanding how systems are 'supposed' to work is legible in everything from city planning to cybernetics to surveillance.

Anthropologist James C. Scott describes how this machinic imagining of the natural and social world drives the powerful to physically impress their machinic imaginings back onto the world:

> The utopian, immanent, and continually frustrated goal of the modern state is to reduce the chaotic, disorderly, constantly changing social reality beneath it to

something more closely resembling the administrative grid of its observations. (Scott, 1998, p. 82)

Scott's work particularly focuses on agriculture and forestry describing this 'utopian' project of reductionism in three stages: first, that the world or field of action (the forest, the farm, the city and so on) is observed and measured, extracting key data points and variables with which to work with. Second, these data points and variables are abstractly modelled, whether in a spreadsheet database, a 'smart' system or a simulation in order to optimize this 'machine' version of reality for the desired outcome. Finally, once a successful model or simulation is devised the world is remodelled to suit it. With the easy accessibility of cheaper sensors and surveillance architectures, more of the world is reshaped to fit the optimal design of a piece of software: Airports, shopping centers, factories and distribution centers are essentially physicalized software – material representations of the way machines imagine the world where 'data' (goods, planes, humans and so on) are processed through them (Kitchin & Dodge, 2011). These and other *Suckling pigs* are non-machine spaces for humans, designed according to a human imagination of machinic worlds that prioritize command and control in search of machinic efficiency.

Sirens

Where the human world is rendered to prioritize legibility for machines.

Automated farms, shipping container ports, airports, pit mines and other software-driven landscapes are edge cases of human and software worlds. These are spaces in which humans and machines move together and as such they bear semblances of legibility and navigability to both (Young, 2019). However, increasingly the world is built in such a way that it is still perceptible to humans but illegible to them. To draw an example that is parallel with Scott's work on industrial forests replanted to prioritize yield for human consumption – a *Suckling pig*, we are already beginning to see forests replanted to prioritize electromagnetic propagation (Manaugh, 2019). These forests, based on science showing the electromagnetic permeability of different species and ideal arrangement and heights, service a physical phenomenon imperceptible to humans and which we can only imagine through the way it instantiates in the physical world.

There are easier examples of human-made worlds for machines: A library of books has a system for both humans and machines to be able to navigate it. The Dewey decimal system allows machines to quickly reference and retrieve items using a system that carries little inherent meaning for most humans without reference to the index. Meanwhile the alphabetical and subject-based organization will allow most humans to find their way to their desired item by act of cross-referencing.

By comparison, Amazon's warehouses or 'fulfilment centers' are almost completely automated and are organized randomly; the most efficient means of rapid storage and retrieval for a machine (Baraniuk, 2015). Here, new goods come in the front and are stored in no particular order. The database managing the warehouse can identify the precise location of each object by a mix of GPS and bar codes but there is no human 'sense' to the order. The imagination at play is completely that of the database. The organization and distribution of objects in these spaces follows the machinic logics and so can often appear possessed by an 'other' intelligence.

Fabulous

Where machines exhibit sublime properties that fuel human imagination.

An often untapped power of machines lies in their ability to broaden the social, spiritual and cultural potential of those who interact with them, with their enchantment of the individual or the society with new possibilities. Anthropologist Alfred Gell describes how mastery of tools and technologies 'enchant' society, projecting power but also a quasi-spiritual and sublime appreciation of non-human artifice (Gell, 1992). In new artificial intelligence technologies such as machine learning, we find ourselves marveling at machines again as they appear to exceed humans with 'processes that are congenitally alien to us—those that are too vast, too distributed or too transtemporal' (Pohflepp, 2016, p. 10).

There is a sublime Lovecraftian sense presented in the contemporary situation in which 'a computer learns and consequently builds its own representation of a classification decision, it does so without regard for human comprehension' (Burrell, 2016, p. 10). Unlike *Stray dogs* this is less because the functioning of the machine is technically obfuscated, forcing us to imagine causality (see also *Having just broken the water pitcher*) but because the very being of a machine itself exceeds human perceptibility. This in turn extends and fuels our own imagination of things previously unimaginable. Deborah Levitt encourages us to use this imaginative opportunity to engage the new technical affordances of new machines of simulation and creativity to dissolve the regime of representation and reality best embodied by *Suckling pigs* (Levitt, 2018, p. 121).

Here we find a great opportunity for 'art'. Projects like Jimmy Loizeau and James Auger's Ripple Counter – part of their 'Sublime Gadgets' series (Auger & Loizeau, 2012) – explore alternative ways that machines might be with us. In this project a floating device simply counts ripples moving across the water's surface. Nearby, a counter with twenty-four digits ticks over towards a septillion, a number that will never be reached. The series uses playful humor and a larger-than-human perspective to playfully draw on philosopher Joseph Addison's notions of the 'Pleasures of the Imagination' in an attempt to de-normalize technologies and explore how they might go beyond or outside the human, revealing to us new perspectives in the sublime. The Slow Inevitable Death of American Muscle by Jonathan Schipper shows us a car crash in excruciating slow motion, teasing the audience with the juxtaposition of power and speed embodied in the powerful American muscle car machines and the imperceptible slowness of the crash spectacle happening over days or even weeks, de-spectaclizing the trope of the car, the car crash and the explosive speed of machines. Finally, the works of Ryoji Ikeda draws the human into a Lovecraftian tale of the profound otherness of the titanic oceans of data produced and disseminated by automated systems. Flashes, explosions, roars of light and sound sweep across enormous spaces without care for human interpretation or presence.

As opposed to *Embalmed*, *Fabulous* do not claim to perform utilitarian technical functions and exist purely to extend the spiritual and aesthetic experience of being human in the presence of machines. These *Fabulous* works and many others like them bring the human under the auspice of the machinic phylum without regard for human ambition, use or purpose and most often beyond the perception of the human senses into the pure otherness of machines.

Stray dogs

> Where humans may struggle to imagine causality or technical function in/of a machine due to their inscrutability. This may give it the appearance of exceeding machine-ness. (aka Wild Machines)

In his book *How Are Things? A Philosophical Experiment* (2003), Roger-Pol Droit describes being overwhelmed by things, by 'folded propositions. Or the folds of ancient and vanished phrases. Or the solid residue of extinct words' (*ibid.*, p. 11). He asks himself 'How are things?' and answers the question himself as an informed imagination:

> The freezer is a machine of secrets. It belongs to the family of thing-enigmas, which perplex us and which we approach with hesitation: the sense of a surface, of volume of a door, of an interior which can be accessed. All of which tells us next to nothing. Things of this kind are self-enclosed, keeping their counsel. Boxes containing mysteries. We become used to them, we draw our own conclusions, but we never really discover how they work. (*ibid.*, p. 115)

Stephen Connor put forward the notion that imagination and machines are inextricably linked because machines seem to exceed their objectness. They can perform tasks and actions that exceed expectations and thus create new imaginary potentials. Mälzel's Chess Player, alternately called the 'Automaton' or the 'Mechanical Turk', was an apparently autonomous chess playing machine constructed by Wolfgang von Kempelen in 1770 which was toured around the courts of Europe for almost fifty years by Johann Nepomuk Mälzel, a Bavarian musician. The machine was an astounding technical marvel, able to beat human opponents such as Maria Theresa and Napoleon I with one commentator describing how 'no exhibition of the kind has ever elicited so general attention as the Chess-Player of Maelzel' (Poe, 1836). This reviewer, Edgar Allen Poe – like others – suspected the machine to be a trick but nonetheless commented that the Chess Player was so astounding a machine precisely because it behaved in a non-machine-like way:

> [A machine's] movements, however complex, are never imagined to be otherwise than finite and determinate. But the case is widely different with the Chess-Player. With him there is no determinate progression. No one move in chess necessarily follows upon any one other. (Poe, 1836)

The non-machine-ness of the Chess Player fueled public imagination precisely because it exceeded the definition of machine. Consequently, dozens of explanations arose for its functioning, some being closer to correct, that the mechanism cleverly concealed a human player directing the machine's movements, and some bordering on the supernatural, that it was haunted by the ghost of a Prussian mercenary. Reports of inscrutable *Stray dogs* exist in the European canon as far back as 807 when the court of Charlemagne received the gift of a mechanical clock from the caliph of Baghdad and continued to fascinate the European world with notions of 'old-world magic' for a thousand years (Truitt, n.d.). However, the Mechanical Turk bears particular weight over the imagination of machines. This is evidenced by the contemporary Amazon's Mechanical Turk *Stray dog*. Here, the inscrutable machine (Amazon) disguises (without apparent irony) a technical functioning that is in fact exploitative human labor.

Included in this present categorization

> Extending from the machinery that produces pasta, plastic buckets, statistics and sound, we find that we consider most things 'machines'. The machine-ness of eating, thinking, and paying attention is directly related to a vague notion of systems. One adds something, and another something is returned. One follows a procedure and a procedure is returned. An action is performed or imagined and something is returned. It is this rhythm of doing and being done that allows the 'Included in this present categorization' category to expand. The category is only defined by its boundary: that which was not included.

The world-as-transactional-machine lets us see the material differently, or rather thinking like a machine makes us look at a field of flax and think of fabric, thread (and muesli). The machine might work by turning a material into another. The flax is ultimately turned into linen and we will look at it and think of thread counts, patterns and moments of use. In this way the machine has turned a field of grass into a material ready for our use and misuse, and it is this turning of material that gives such a machine its identity and power. By turning a material into something else we engage in the machine magic of material transformations.

Frenzied

Where machines propagate new imaginaries.

* This may be read as a oppositional categorization to things Belonging to the emperor in the sense that Frenzied necessarily challenges status-quo imaginaries and acts against imaginary foreclosure. Further research is needed to study machines or imaginings that may cross both categories, perhaps necessitating a new category.

Machines also come with imaginable futures attached to them: from Gutenberg's Press and the Protestant revolutions to the car and the twentieth-century version of freedom. Machines are entangled and drawn up in the social imagination of what our world is and what we want from it. At the individual level, the latest gadget or gizmo comes not with a technical specification but a set of promises and visions of what life will be like once one holds it, puts it in one's pocket and uses it.

These become most revealed in deviant edge cases, where machines serve new purposes or are adapted. Here we find the site of Jugaad innovation, the reclaiming of machines for new purposes as well as hacking, exploitation and deviant misuse. In the video game subculture of speed-running, for example, the game is read as a digital architecture through which new games can be created using glitches and inconsistencies in the design to deceive the developers' intended purposes. This playful approach to the machines of the game relies on a canny understanding of its affordances and limitations only discovered through experimentation. More sinisterly, the use of the burgeoning Internet of Things ecosystem as a base from which to launch distributed denial of service (DDoS) and ransomware attacks has shown how the technical attributes of machines will always allow for those who wish to exploit them.

The *Frenzied* machine buzzes with the promise and potential of itself, but it is a broken promise, reminiscent of the broken promise of the souvenir that promises a connection that it cannot deliver (Stewart, 1993). The *Frenzied* machine is frantically insistent on continuing this non-delivery, as a culturally complex DoS attack, ready to be subverted to the purposes of art and harassment.

Innumerable

Those machines and imaginings we have forgotten.

Alfred Gell describes technology as made of three things: the sum of the 'artefacts which are employed as tools ... the knowledge which makes possible the invention, making and use of tools [and] the networks of social relationships which ... provide the necessary conditions' for their use (1988, p. 6). In other words, a machine is dependent on its discrete existence, the knowledge of its use and the social necessity of its use. Innumerable are the machines that have fallen out of knowledge, necessity or existence and become forgotten. Certainly, too many to recount here.

Drawn with a very fine camelhair brush

Where machines render imagined things.

Equally, machines shape how we imagine. The growth of computer-generated image technology and attendant platforms like virtual, augmented, extended and mixed reality have allowed us to visualize and share imagination in new ways. Here we find the dreams of simulation and holography, the 'Holodeck' of Star Trek and the Deep Fake. These machines imagine worlds for the human senses. However, in doing so they shape those worlds in accordance with their software predispositions. As Alan Warburton points out, 'Computer scientists incrementally created a library of simulated phenomena [and] software companies packaged these tools together into multi-purpose 3D animation programs. These creative suites naturally prioritize certain tasks and outputs. They ship with presets for lights, objects, motions, bodies and materials' (Warburton, 2019).

The machines used to construct visions of the future, whether concept products, renderings of future developments, science-fiction cinema or data visualization, increasingly rely on standardized software packages with standardized templates. As a consequence, imagination of the world and the future becomes limited and pre-described, homogenized by the software used to envision it. As Joel McKim suggests, 'Memory is becoming increasingly homogenized through both the conditioning or standardization of user-generated material and the perpetual re-circulation of a relatively small (and increasingly commercial, rather than amateur) pool of available content' (2017, p. 291).

The power of *Drawn with a very fine camelhair brush* is in the excellence and fidelity of their production of imagination: the advances in real-time photorealism, facial recognition, machine vision and deep learning that have enabled more and more granular simulation of natural phenomena. As a consequence, they can be used for disinformation or exploitation: 'Computational propaganda has become a normal part of the digital public sphere ...These techniques will also continue to evolve as new technologies ... are poised to fundamentally reshape society and politics' (Stubbs, 2019). Recent elections have seen the proliferation of doctored videos and deep fakes to sway political opinion, often drawing on and confirming what a viewer already imagines to be true.

Et cetera

> All the things not included in the 'Included in this present categorization' category. Also known as 'the rest'.

Taken out of any sensible or coherent context, a well-worn Whitehead quote reads, 'A traveler who has lost his way should not ask, where am I? What he wants to know is, where are the other places' (Whitehead, 1978, p. 170). *Et cetera* are these other places. In a world of machines, we may find ourselves looking for the not-machines, that which is leftover or material not yet constructed into machines. We may consider *Et cetera* the category for material in what Annie Albers would call its original form: unconstructed and unprocessed (Albers, 1937).

Having just broken the water pitcher

> Where previously unimagined machines or uses of machines may emerge serendipitously. Maybe.

While imagining future technologies through the notions of machine-ness might be strongly influenced by what we already imagine to be possible (see *Belonging to the emperor*), sometimes machines defy our imagination. Here, they can become alternately alienating or inspiring. We are shocked and displaced when machines behave in ways we could not have previously imagined. Popular media is replete with stories of technology behaving in new or unexpected ways, whether by design or accident. With an increasing amount of 'smart' things in the world, we are increasingly facing unexpected surprises in how we imagine our life with these things to be. The story of the smart lock that would occasionally not lock due to a bug (Brown, 2014) or Amazon Echos laughing maniacally on their own (ITV, 2018) indicate a world of perceived autonomy beyond our imagination.

These happenings often drive us to imagine supernatural explanations for machines that are already presented as somewhat magical. Further to this, the black boxing of these technologies in their complexity and legal and technical abstraction force us to dig for primal explanations of their functioning and glitching.

William Stahl (1995) in 'Venerating the black box' draws on the similarities in the narratives of power around technology and the occult that lead to 36 per cent of TIME magazine articles about the PC referring to it and its creators in terms of the occult. In his work he demonstrates how magic and technology are entwined in the pursuit of power with the resulting superhero fantasies that pervade the media reporting on machines.

This magical attribution falls on an apparent lack of perceptible causality in the operation of machines and their inscrutability (see *Stray dogs*). In our need to imagine the functioning of these machines we will often create new imaginaries or ideas that propagate ad infinitum to create new machines and imaginations. Apophenia, or the tendency to connect unrelated things in the imagination, can in fact lead to imagination. Hito Steyerl writes that 'apophenia happens when narrative breaks down and causality has to be recognized—or invented—across a cacophony of spam, spin, fake, and gadget chatter' (2018, p. 5). Here, apophenia results in invention: creating new, previously unimaginable meanings through accident as when the messy outcome of an overtrained neural network 'reveals its hard-wired ideologies and preferences' (*ibid.*, p. 10).

That from a long way off look like flies

Where machines become props, tools or partners in imagining new worlds.

The relationship between the machine and the imaginary is recursive. For example, the car enables us to imagine the flying car, the electric car, the autonomous car and others. In this way, new machines are prefigured by the imagination of the previous machines. Without the first car, the autonomous car would have been unimaginable. Broadly speaking, we imagine new machines with the machines we have, and the machines we have define the things we imagine. To misquote and paraphrase: We imagine our machines; thereafter, they imagine us.

As opposed to *Embalmed* that are only even imaginary in existence or function, *That from a long way off look like flies* cross the imaginary/real divide. They are speculations and distant proposals for new machines and imaginations. They promise new futures and better worlds. One might find an iPad and consider its similarity to the devices proposed in *2001: A Space Odyssey* or marvel at how the countdown clock of Fritz Lang's *Woman in the Moon* became a central motif of space flight. These machines usher in nascent realities as portals on the edge of the imagination and reality.

From here, *That from a long way off look like flies* might progress into other categories, passing out of reality and into the purely imaginary as *Embalmed*. Or they might find serendipitous use as *Having just broken the water pitcher*. Or exceed human perception as *Fabulous*.

We have tried to describe a world of imagined and actual machines. The machine-ness of our surroundings is both a description of the fascination and horror with that which we do not understand or are barred from touching and an extension of a perception of everyday life as procedural and sometimes barely functional. We approach each machine with a whispered curse or plea: please work, please don't harm me, please respond. We are deeply emotionally bound to the success of these encounters, we modify what we want to what the machine can do and we cover for its errors. Entire careers are given over to the cleaning up of the mess from the machines.

The absurdity of categorizing the entangled worlds of humans and machines and their co-imaginative potential should by now be absolutely apparent. Humans are technopolitical beings and construct machines into their social imagination. In attempting to categorize them we draw attention to a paradox that would have pleased Borges: No matter how absurd the categorization, we are providing proof of its validity by reflexively imagining with machines in the mere acts of researching, recalling, writing, editing, talking about and reading this very chapter.

Bibliography

Albers, A. (1971). *On designing*. Middletown: Wesleyan Press.

Auger, J. & Loizeau, J., (2012). *Sublime Gadgets* [Art]. Retrieved 12 March 2021 from http://www.auger-loizeau.com/projects/sublime-gadgets.

Baraniuk, C. (2015). How algorithms run Amazon's warehouses. *BBC Future*. Retreived 25 September 2019 from http://www.bbc.com/future/story/20150818-how-algorithms-run-amazons-warehouses.

Bassett, C., Steinmueller, E., & Voss, G., (2013). *Better made up: The mutual influence of science fiction and innovation*. Nesta working paper.

Brown, J. (2014, October). Review: August smart lock. *Wired*. Retrieved 23 September 2019 from https://www.wired.com/2014/10/august-smart-lock-review/.

Borges, J. L. (1952). The analytical language of John Wilkins. *Alamut: Bastion of Peace and Information*. Retrieved 19 April 2020 from https://ccrma.stanford.edu/courses/155/assignment/ex1/Borges.pdf.

Burrell, J. (2016). How the machine 'thinks': Understanding opacity in machine learning algorithms. *Big Data and Society*, *3*(1), 1–12.

Connor, S. (2017). *Dream machines*. London: Open Humanities Press.

Daston, L., & Galison, P. (2010). *Objectivity*. Cambridge, MA: Zone Books.

Deleuze, G. (1988). *Foucault*. Minneapolis: University of Minnesota Press.

Droit, R-P. (2003). *How are things? A philosophical experiment*. Translated from *Dernières nouvelles des choses*. London: Faber and Faber.

Enns, A. (2019). Apocryphal psychotechnologies. In J. Allen & A. Enns (Eds), *Continent*, 8(1–2). Retrieved 16 April 2020 from http://continentcontinent.cc/index.php/continent.

Gell, A. (1988). Technology and magic. *Anthropology Today*, *4*(2), 6–9.

Gell, A. (1992). The technology of enchantment and the enchantment of technology. In J. Coote & A. Shelton (Eds), *Anthropology, Art and Aesthetics* (pp. 40–66). Oxford: Clarendon.

Graeber, D. (2015). *The utopia of rules: On technology, stupidity and the secret joys of bureaucracy*. Brooklyn: Melville House.

Hayles, N. K. (2019). Can computers create meanings? A cyber/bio/semiotic perspective. *Critical Inquiry*, *46*(1), 32–55.

ITV (2018, 8 March). Amazon Alexa users spooked by creepy 'laugh' emitted by their devices seemingly at random. *ITV News*. Retrieved 25 September 2019 from http://www.itv.com/news/2018-03-08/amazon-alexa-users-spooked-by-creepy-laugh-emitted-by-their-devices-seemingly-at-random/.

Kitchin, R., & Dodge, M. (2011). *Code/space: Software and everyday life*. Cambridge, MA: MIT Press.

Latour, B. (2000). *Pandora's hope: Essays on the reality of science studies*. Cambridge, MA: Harvard University Press.

Levitt, D. (2018). *The animatic apparatus; Animation, vitality and the futures of the image*. Alresford, Hampshire: Zero Books.

Losh, E. (2016). Sensing exigence; A rhetoric for smart objects. *Computational Culture*, *1*(5). Retrieved 12 March 2021 from http://computationalculture.net/sensing-exigence-a-rhetoric-for-smart-objects/.

Manaugh, G. (2019). Computational landscape architecture, *BLDGBLOG*. Retrieved 19 April 2020 from http://www.bldgblog.com/2019/02/computational-landscape-architecture/.

McKim, J. (2017). Speculative animation: Digital projections of urban past and future. *Animation: An Interdisciplinary Journal*, *12*(3), 287–305.

Merriam-Webster (n.d.). Machine. *Merriam-Webster's Dictionary*. Retrieved 24 September 2019 from https://www.merriam-webster.com/dictionary/machine.

Poe, E. A. (1836). Maelzel's Chess-Player. *Southern Literary Messenger*, *2*, 318–326.

Pohflepp, S. (2016). *Pattern agnosia and the image not made by human hand*. Retrieved 12 March 2021 from https://www.academia.edu/30482204/Pattern_Agnosia_and_the_Image_not_made_by_Human_Hand.

Scott, J. C. (1998). *Seeing like a state: How certain schemes to improve the human condition have failed*. New Haven, CT: Yale University Press.

Singleton, B. (2014). *On craft and being crafty: Human behaviour as the object of design*. PhD thesis. University of Northumbria, Newcastle.

Stahl, W. A. (1995). Venerating the black box: Magic in media discourse on technology. *Science, Technology, & Human Values*, *20*(2), 234–258.

Stewart, S. (1993). *On longing: Narratives of the miniature, the gigantic, the souvenir, the collection*. Durham, NC: Duke University Press.

Steyerl, H. (2018). A sea of data: Pattern recognition and corporate animism (forked version). In G. Bachman, T. Beyes, M. Bunz & W. H. K. Chun (Eds), *Pattern Discrimination* (pp. 1–22). Minneapolis: University of Minnesota Press.

Stubbs, J. (2019). Viral visuals driving social media manipulation on YouTube, Instagram: Researchers. *Reuters*. Retrieved 29 September 2019 from https://www.reuters.com/article/us-facebook-disinformation/viral-visuals-driving-social-media-manipulation-on-youtube-instagram-researchers-idUSKBN1WB0ED.

Truitt, E. R., (n.d.). Preternatural machines. *Aeon*. Retrieved 19 April 2020 from https://aeon.co/essays/medieval-technology-indistinguishable-from-magic.

Warburton, A., (2019). Fairytales of motion [Video]. Retrieved 30 April 2020 from https://vimeo.com/343626951.

Ware, C. (2001). Superpowers. *This American Life* [Podcast]. Retrieved 30 April 2020 from https://www.thisamericanlife.org/178/superpowers.
Whitehead, A. N. (1978). *Process and reality*. New York: The Free Press.
Young, L. (2019). *Machine landscapes: Architecture of the post-anthropocene*. Chichester: John Wiley & Sons.

PART TWO

INTERACTIONS

4 WHAT ARE YOU? NEGOTIATING RELATIONSHIPS WITH SMART THINGS IN INTRA-ACTION

Christopher Frauenberger

We are deeply entangled with and implicated in the material world around us. While historically the Enlightenment has led us to think that mind and matter are strictly separated, postmodern thinking, in particular in the phenomenological tradition, has sought to bring these dualisms down. The embodied understanding of our tool use, most prominently Heidegger's hammer, was compelling, but it is now that we are surrounded, immersed and embedded in smart, interactive technologies that this intimate entanglement is gaining new qualities which require us to evolve our ways of thinking about the relationships with our (smart) tools. So, this chapter aims to do precisely that; it seeks to develop an argument for re-conceptualizing our relationships with smart everyday things with implications for their design and our use of them.

There are three main reasons, I argue, that should motivate us to evolve our thinking about the relationships we have with smart everyday objects. First, the pervasiveness of smart and connected things in our lives is on a steep rise, reaching into almost every corner of modern life. It is anticipated that by 2030, there will be more than 125 billion devices connected to the internet (HIS Markit, 2017), some smarter than others. The extent to which this shapes who we are and what we do makes it paramount to overcome the notion that smart things are just clever tools. This relates to the thinking of the German-Austrian philosopher Günther Anders (see Müller, 2016, for an English translation) and later Bernard Stiegler (1998) who have looked to Greek mythology to make sense of us and technology. According to the myth, Epimetheus, the not-so-bright brother of Prometheus, was tasked to distribute traits among all living beings, but eventually ran out of them before he got to humans. Thus, humans were born naked, but Prometheus, against the will of the gods, decided to bring humankind fire instead and kick-started civilization, the arts and the sciences. While Prometheus was cruelly punished for his crime, the story provides a productive lens to think about the false opposition between humanity and technology. Anders harnesses the Promethean myth to argue that humans are fundamentally alien to the world ('Weldfremdheit') and only come to be part of it by defining themselves through their use of the material world. In other words, we are in very fundamental ways defined by what we build for ourselves. What we create shapes who we are in the world. Certainly, in the days of smart everythings and us, this rings truer than ever. Mobile digital technology has already made us very different kinds of people and societies, and we are yet to find out what the increasing pervasiveness and smartness of machines will lead us to be in the world.

A second quality of smart everyday things heighten the stakes: their chameleonesque capacity to evolve what they are, not only over generations of things, but also within their lifetimes. While a hammer is a hammer and relatively stable in its material manifestation and cultural interpretation, artificial intelligence (AI) and unsupervised learning techniques have opened up possibilities for machines to truly develop and become other things without any direct intervention of us humans. We can think of the many AIs that turned racist when interacting with the world (a well-known example is Microsoft's AI called Tay that had to be switched off within 24 hours of interacting through and learning from Twitter). Further, in many cases, we simply cannot know about the inner workings of an AI and how it comes to reacting in certain ways for certain data. We therefore cannot know about how they might evolve provided with unforeseen input. The malleability and the ontological volatility of smart things in use adds to complicated questions around accountability and highlights the need to think whether any meaningful difference can be made between design time and use time of smart everyday things.

Third, and following from the above point, the 'smartness' of things has the potential to express intent in ways few not-smart objects can. The agenda or intent one might embed in an algorithm can be both more tacit and more powerful and consequential than that one could embed in a hammer. As such, their claim to agency is much stronger and their potential to contribute to, guide or manipulate human activity is much greater. Let us just think about the modern breed of speech assistants: Heather Woods (2018), for example, analyses how Alexa and Siri invoke a feminine persona that transport normative stereotypes of a caregiver, mother and wife in service of modern surveillance capitalism (Zuboff, 2019) – that is, our anthropomorphizing of these devices is harnessed to manipulate us to consume (or vote) the right way. As a side effect of this agenda, we are seeing users' interactional style change beyond their conversations with the speech assistant. Bonfert *et al.* (2018), for example, investigate how to design AIs that rebuke impolite voice commands to address the problem of children adopting rude language from their interactions with speech assistants as an acceptable norm. Again, this level of impact on human activity raises serious questions about accountability, responsibility and the ethical implications of building and using such technology.

To tackle these challenges, I propose to look to a range of thinkers who have developed relational and performative concepts as a way to overcome the social-natural and cultural-material dualisms which, I argue, are hindering us to appreciate and account for the deep mutual interdependencies between humans and machines. Taking the work of Karen Barad and Bruno Latour, among others, as a starting point and inspiration, I aim to draw out what relational ontologies can do to help us understand the role of things in human activity. I first will provide a selective review of relevant theory before developing more concretely a theoretical position on the nature of relationships between humans and smart things. In the fourth section I then discuss what such a position suggests for designing and using smart objects. This leads me to argue for understanding both, the design and the use of smart things as a process of continuous negotiation for which we have to create and maintain appropriate spaces. Drawing on the work of political philosopher Chantal Mouffe, I conceptualize these spaces as agonistic arenas in which we can have constructive conflicts over agency, power and morality with smart objects. I advocate for a new kind of smartness in a new breed of things that allow us to negotiate and enact desirable technological futures.

Entanglement theories

The three lines of thought briefly reviewed here – Actor-Network Theory (ANT), Post-Phenomenology (PP) and Agential Realism (AR) – are all part of a larger movement towards overcoming the dualisms that were entrenched by modernity. In particular they seek to abolish the notion that there is a social realm in which matter has no role, other than acting as the passive backdrop to human intentionality. Entanglement theories, a label I derive from Wanda Orlikowski (2010), all start from the premise that matter plays a central role in configuring social life, that is, it contributes to the configuration of agency. Or as Lucas Introna (2014) puts it, such a perspective 'posits the social and technical as ontologically inseparable from the start'. Such decentring of the human in favour of acknowledging the role of the material world in a relational ontology has also led to such theories being labelled as new materialism or post-humanism.

Putting humans and non-humans onto the same level is far from uncontroversial and typically polarizes any audience. Particularly in times when we seem to finally make headway in shifting technologists from a because-we-can mentality to a more user-centred approach. However, as I will argue below, counter-intuitively maybe by working to make the active role of technology visible in configuring human activity, we gain handles on accountability and the ethical dimensions of our work.

The following is a purposefully selected review of some of the many theories that could be called entanglement theories. They are necessarily brief and reductive as they are only intended to provide enough background to underpin the line of argument developed below. I provide a more elaborate discussion on entanglement theories in human-computer interaction in Frauenberger (2019) and encourage readers to delve into the wealth of related literature, for which the references here can only be a starting point.

Actor-Network Theory

Latour and Collin developed ANT within the context of Science and Technology Studies (STS), aiming to understand the ways in which we conduct science within social as well as material surroundings. Importantly, it is distinct from a purely social constructivist account in that it not only acknowledges the sociocultural context in constructing knowledge but also recognizes the key role of the material world. This leads them to argue for a perspective in which human and non-human actors are on an equal footing, arranged in a network of associations which determine the activities enacted. Agency, power, knowledge or experience consequently cannot be localized within actors but are determined as an effect of the actor network.

Latour (2005) begins his argument with a direct attack on classical sociology by questioning how we ended up separating the social from everything else. He argues that there are no 'social aspects' of otherwise non-social things in the world, but everything (humans and non-humans) belongs into the same realm. All actors and their associations make up our reality. He therefore also calls ANT a sociology of associations in contrast to the old sociology of the social. Material (and smart) things allow, afford, restrict or enable human activity; they are themselves sources of action and become participants in the course of this action. Things do not determine what humans do, as the materialism in Marx could be interpreted, but they are also not the inconsequential, passive backdrop for humanity, as suggested by the sociology of the social. As Latour puts it, there exist 'many metaphysical shades between full causality and sheer inexistence' (*ibid.*, p. 72), so ANT is interested in the many different levels in between at which things participate in action. And ANT goes further and argues that anything that has influences on an action in a way described above is an associated actor, which also includes non-material entities such as policies, laws or societal norms.

Post-phenomenology

Unsurprisingly, PP starts in a very different corner, but it ends up in a very similar place as ANT, with a relational ontology that foregrounds the role of the material world. Don Ihde (1990) develops the fundamental ideas of Martin Heidegger's *Dasein* (being-in-the-world) further to account for the expanding relationships with our tools. Mediation is the concept by which he elevates the role of things in actively relating our *Dasein* to the world. In other words, our being-in-the-world is always mediated by technology. Ihde (1990) distinguishes between different kinds of relations, for example, embodied (technologies extend the body for certain actions), hermeneutic (technologies extend the natural world for certain actions), alterity (explicit interaction) and background (technologies as context for human existence) relations. Peter-Paul Verbeek (2008) expands these to include various cyborg relations that can be immersive or augmentative.

Importantly, PP develops intentionality further to go beyond mere relationality between thoughts and things. It argues that subject and object are constituted in their mediated relation – that is, any human experience of the world is mediated by material entities, and both humans and non-human entities define themselves through this relation. 'Human-world relations are practically enacted via technologies' (Rosenberger & Verbeek, 2015, p. 12). As such, PP also subscribes to a performative, rather than static, ontology. In other words, entities are not predetermined but are defined in action through their relation to others. Heidegger's famous hammer or the blind man's cane that Maurice Merleau-Ponty discusses are not fixed entities, but as they mediate people's experience of the world, the boundaries between them and their users are enacted differently, depending on the activity.

This also leads to a position that is reminiscent to the arguments brought forward by Anders: Whoever creates technologies that we interact with contributes to defining who we are. Arguing for mediation theory as a viable perspective on interaction design, Verbeek (2015, p. 31) posits this very clearly: 'Designing technology is designing human beings.' As an example, he brings up social media and how it has brought about new 'types and dimensions of social relations' and has thereby contributed to 'shape human existence'.

Agential realism

In her book *Meeting the Universe Halfway*, quantum physicist and feminist philosopher Barad (2007) develops her own theory to conceptualize the entanglement between humans and matter – AR. Her starting point is the philosophy-physics of Niels Bohr and the double-slit experiment that demonstrates the particle-wave duality of light. Depending on what is measured, either the distribution of light behind the two slits or through which slit light is travelling, light either behaves like a wave producing the usual interference patterns or as particle, producing a very different distribution on the screen behind the two slits. For Barad, this shows how the material configuration of measuring changes the ontological nature and boundaries of entities. In fact, she argues that reality comes about through the production of phenomena by the intra-action between entities in a certain configuration, that is, human and non-human actors continuously enact reality that is causally linked to the configuration that is present. Importantly, Barad resolutely rejects the relativism of postmodern thought: Reality is not arbitrarily constructed in the social realm but is causally determined by the configuration of actors (material or human). This is also why she sees her theory as a new form of realism.

Drawing on feminist scholar Judith Butler, Barad (2007) also shifts her ontology from static representations to performativity. She argues that the boundaries between entities are drawn in their intra-action,[1] so, where we end and some smart object starts may only be decided once we start using it. AR introduces uncertainty over the absolute completeness of ontological entities. Again, Merleau-Ponty's discussion of a blind man's cane serves as an example here. As such, AR is also decidedly post-human. It decentres humans and distributes agency and accountability across all actors, humans and non-humans.

Summarizing, AR has the following cornerstones:

1. The primary ontological unit of reality is not bounded entities but phenomena that are reliably (and objectively) (re)produced by discursive material practices – something Barad calls mattering.
2. Things and people, as phenomena, mutually constitute each other through their intra-action, that is, the boundaries between humans and machines are not predetermined but enacted by making agential cuts.
3. What is possible to enact depends on the material configurations, that is, reality is causally produced through a certain intra-action within human and material configurations. This allows her to trace responsibility within these configurations with rigor.
4. The world is in an open-ended and continuous process of mattering, that is, these configurations constantly change and produce different agential cuts and phenomena.

[1] Barad coins the term 'intra-action' precisely to emphasize that two entities are not merely interacting with each other but defining each other through their intra-action.

We and smart things

The above highlights only the most salient features of different entanglement theories and aims to provide the bare minimum for underpinning the perspective on the relation between humans and their smart things I aim to develop here. To ground my argumentation, I will make use of a running example to exemplify some of the aspects in the relationship between humans and their smart, everyday objects: a hypothetical, smart fitness tracker.

It may be less of a radical proposition to say that smart things are sometimes of different kinds for different people. For some, a fitness tracker really is just a watch that gives the time. But the notion of **mutual constituency** that is inherent in all of the theories above opens us to the view that it is not only us defining smart things but also vice-versa. In our intra-action with smart things, agential cuts are being made and certain phenomena are enacted that are causally linked to the configuration (i.e. relations) at hand. During exercise, a smart fitness tracker may become embodied, an integrative part of me regulating my heart rate in action, for example. At home, and as part of a different configuration, a fitness tracker becomes an object of surveillance that provides performance data to a platform that has an agenda of behaviour change through praising or shaming, maybe by leveraging peer pressure. On a higher level, a fitness tracker may also more fundamentally change who we are by linking aspects of our self-image and happiness to whatever the fitness tracker can measure and the platform may classify as desired behaviour. Further, smart fitness trackers change their behaviour too, resulting in a perpetual circle of self-reinforcement of inscribed, normative behaviour. Katta Spiel (2019, p. 35) reflects on fitness trackers and body positivity and puts it this way:

> Not only does the tracker learn to refine its judgements by having more data from the human, but the human also learns how to perform certain activities so they are judged as relevant by the tracker. Together, the person and the technology establish a coherent reality that neither could have created by themselves.

So, using a smart thing is essentially also a process of figuring out whether one wants to be what the smart thing is making one to be.

Further, the entanglement between humans and their (smart) things is constantly reconfigured and continuously enacted. The **performativity** of the boundary-making and entailed ontological uncertainty means that meanings and relationships keep shifting on small and large time frames. Smart fitness trackers may be tools, helpers, friends, hate objects or completely disappear in the mediation of our being-in-the-world. The enacted phenomena may change quickly depending on the activity or shift slowly as the novelty effect of making training progress visible is wearing off. According to a 2014 report, about one-third of users abandon their tracking devices within three months (Ledger & McCaffrey, 2014), and Clawson, Pater, Miller, Mynatt and Mamykina (2015) found the most frequently stated reason relating to a mismatch of expectations. In their study, Clawson *et al.* (2015) also found related evolving practices, in which the use of fitness trackers changed with circumstances, in response to the messy realities of lives or an increased own understanding of what users want. Entanglement theories provide a very effective lens to understanding these shifts as reconfigurations that enact different phenomena.

This example also emphasizes the **networky** nature (Latour, 2005) of the socio-material system at play. It is obvious that fitness trackers and users are not the only actors here. The larger actor-network may include other people (your doctor, your spouse, your peer group), other things (your running shoes), other smart things (your mobile phone, the self-tracking platform) and other non-human things (societal norms, the General Data Protection Regulation, the sales pitches) – all of which have some materially discursive properties (Barad, 2007), programmes of action (Latour, 2005), mediating qualities (Ihde, 1990) or a dispositif (Foucault & Gordon, 1980). In other words, they all contribute to the phenomena enacted in specific ways that causally link the configuration to the action. Introducing a new smart thing is reconfiguring this actor-network and there is no way to isolate any intra-actions from the rest of the network.

At this point it is important to discuss the nature of **agency** from the perspective of entanglement theories. Agency is no quality that anything or anyone can possess, but is an effect created by the materially discursive configurations of human and non-human actors. In other words, neither the user nor the fitness tracker possesses agency in any static way or form. Agency is enacted in their intra-action and distributed in the network, that is, how much anyone's or anything's programs of action influences the resulting action is dependent on the configuration of associations. So, how much intended behaviour change a fitness tracker causes is dependent on how it is configured in the network of other actors.

Which leads us to the question of **power** and **responsibility**: who can (re)configure networks of actors? Entanglement theories would swiftly reject the claim that it is entirely in the power of humans to lay out configurations. The material and more generally the non-human world allows, affords, restricts or enables phenomena in ways that is well beyond the influence of human intentions. This is not to say that design decisions or use behaviours do not matter. Quite to the contrary, entanglement theories posit that human intentionality cannot determine phenomena on their own, but they also hold all actors accountable for their contribution to shaping action. With smart things and their smart infrastructures, the possibilities of shaping phenomena is rather significant; see the grip surveillance capitalism has on democracies and what the role of technological platforms is in this (Zuboff, 2019). Tracing agential cuts and programs of action in associations then allows us to create accountability within the network of actors, even if some of them are non-human (e.g. a societal norm or an AI that went off the rails).

Consequently, designing smart things should be conceptualized as the process of reconfiguring actor networks, producing new potentialities for enactments. The intent or inscribed program of action reflected in design decisions matters to the possibilities of enactments that are suggested. To quote Kranzberg's first law of technology, 'Technology is neither good nor bad; nor is it neutral' – our fitness tracker can offer potentialities that lend themselves to actions of performance optimization, consumerism, losing weight, body positivity, self-determination and so forth (Kranzberg, 1986). Or, we can think of it as something that may be an object with intent but is malleable and open to negotiate its material personality as we begin our relationship and try to find out what we want it (and us) to be. The following aims to expand on this notion and discuss what this might mean for the design-use of smart things.

Design-use of smart things

The extent to which smart things are entangled with us, how our relationships with them shift in intra-action and are reconfigured over time, has two important implications for design: (1) We need to find ways in which people can participate in designing meaningful relationships and (2) we need to find ways for how this participation can continue across design *and* use times. So, the question is not only how we design smart objects with intent (see Rozendaal, Boon & Kaptelinin, 2019, for a discussion) but also how we configure participation around the design-use of these smart things. To this end, I argue to look to the field of participatory design (PD) and its interpretation of concepts such as infrastructuring, meta-design and agonism.

PD originated in the 1970s in Scandinavia and the UK and was initially concerned about the future of work. Strong unions recognized that this future was intimately tied to the roles technology would play. While capitalism had a rather straightforward idea about what technology should be doing, namely reducing costs by replacing humans, the unions were interested in developing an alternative vision that made workers part of that future (Nygaard & Terje Bergo, 1975; Ehn, 1989). Since then, the field, now best characterized by the community around its biannual academic conference, has tremendously diversified in terms of contexts, methodology and motivations. Michael Muller (2003) provides an early overview of the many different methods that the field has produced to involve a wide variety of stakeholders, including the users, in the design of technological artefacts. Finn Kensing and Jeanette Blomberg (1998) point out that the motivations to work in participatory ways may best be described as a spectrum from pragmatic to idealistic. While pragmatic PD work focusing to leverage user participation to maximize the system-user fit, the ideological end of the spectrum emphasizes the democratization of creating technological futures and the empowerment of users. Whatever the orientation, what PD has provided us with is a well-tested toolbox of approaches that allow us to meaningfully involve people in design processes.

However, PD is not without its challenges, most prominently maybe the issue of idiosyncrasy and scale (Frauenberger, Foth & Fitzpatrick, 2018). Much of the work in PD is situated within a limited, well-defined scope and time frame. This makes it hard to transfer processes from one context to another, let alone scale them into market places. Several concepts have been discussed to tackle this challenge, most notably the notion of **infrastructuring**. Originating in the work by Susan Star and Karen Ruhleder (1994), PD has picked up on the socio-material emphasis and the idea that infrastructured technologies can be interpreted and used in different ways by different users at different times (e.g. Karasti, 2014). Pelle Ehn (2008) has made similar arguments in extending the notion of a design thing to mean the actual object within its socio-material and sociocultural context. In many ways, this perspective is inspired by and resonates with Latour (2005).

Another approach to tackle issues of idiosyncrasy, scope and scale that Ehn (2008) brings up is **meta-design** – the idea of designing things for their continuous design in use. Coined as a term by Gerhard Fischer and Elisa Giaccardi (2004), Meta-design recognizes that the envisioned use of technological artefacts can be quite different to the actual use and argues for providing spaces for the user to develop/appropriate/customize the design to their needs. Designing for this possibility is what they call meta-design. As Ehn (2008) points out, this has a number of strategic consequences for PD. It means that future users need to be considered stakeholders in the process, without them being involved in the design games[2] during the design phase. Again, this points to not only creating artefacts but also infrastructures, environments and platforms for future users to participate in the design-use games.

I argue that entanglement theories provide the ideal theoretical underpinnings to conceptualize this kind of continuous design-use within its socio-material context. It recognizes not only future users as actors in the meaning and boundary making but also future material contexts. In the ongoing reconfiguration of all human and non-human actors and their intra-actions, the boundaries between technologies and humans are (re)drawn, meaning and activity is (re)constructed, or better, negotiated. From this perspective, design and use times are effectively blurred: Designers and stakeholders intra-act with material affordances to negotiate a thing with intent (Rozendaal *et al.*, 2019). In the *making* they configure

[2] Ehn (1989) derives this term from Ludwig Wittgenstein's language games to signify the negotiation of meaning through the activity of designing together, rather than using words.

themselves, stakeholders and future users around a matter of concern (see Latour, 2005); however, design does not end then. In use, the thing is continuously reconfiguring the actor-networks of people, creating meaning and affording activities.

In other words, rather than thinking of 'users of things', we should think of people who continuously negotiate their relationships and boundaries with smart things as co-designers, collaborators and co-habitants. This notion goes well beyond getting used to or customizing smart things or 'breaking them in'. Taking a relational ontology perspective offers to describe our relationships with smart things as a process of defining and continuously redefining us and the smart thing through our intra-action. This reorientates designing smart things from the magic-like tool towards an actor with intent that needs to be open to negotiate its relationship in a wider network of use. Maybe, it is this responsiveness to intra-action in the world that makes future smart things truly smart.

It is important to recognize that all intra-action reconfigures agency and thus power. All human and non-humans actors come with intent, a program of action, a dispositive, an affordance or offer a mere possibility that enables, shapes or restricts the intra-action. Further, by configuring and reconfiguring actors, the way and extent they contribute to the intra-action changes. There might be less scope for negotiating the affordances of a hammer at use time and its contribution to action, if this is primarily defined by how its use is configured in relation to other actors, for example, nails, a piece of wood, a carpenter. However, it is easy to see that both the program of action and the configuration that determines the contribution of a smart thing to any intra-action could be made more accessible to people while using it. What a fitness tracker is and tries to do in relation to us and who we become in return can be made subject of a continuous negotiation.

Such negotiations are by no means innocent; they are not rational or a form of objective decision making. They are political activities in which agendas and power differences are the main drivers of change. The fitness tracker may be designed to provide its functionality only when used in the context of competitive comparison – this is a materially discursive power grab. People may fight back and subvert programs of action, for example, by only ever using a fitness tracker without connecting it to the internet. I argue that these spaces of negotiation are poorly understood, if not completely ignored by technologists. As designers, the intended use is what we work for; everything else is accidental and a failure.

To better conceptualize and work with these spaces for negotiation in design, PD has looked to the work by the political philosopher Chantal Mouffe (2013) and her concept of **agonism**. She argues that conflict and controversies are the raison d'être of politics and that the struggle for hegemony is nothing politics needs or should solve by striving only for consensus. Rather, she argues for creating spaces in which conflict is nurtured but framed as agonism between adversaries, that is, a vigorous, but non-violent, respectful and constructive struggle rather than antagonism between enemies that require dominance and oppression as the outcome. The introduction of these political concepts in the field of PD led to the notion of agonistic participatory design. Björgvinsson, Ehn and Hillgren (2012, p. 143) describes the shift as follows:

> The design researcher role becomes one of infrastructuring agonistic public spaces mainly by facilitating the careful building of arenas consisting of heterogeneous participants, legitimizing those marginalized, maintaining network constellations, and leaving behind repertoires of how to organize socio-materially when conducting innovative transformations.

In other words, agonistic participatory design creates arenas to facilitate constructive conflicts in materially discursive negotiations between actors about their programs of actions and the configuration of actor-networks. Again, such an understanding of the role of design further weakens any meaningful distinction between design and use times. In Frauenberger, Spiel, Scheepmaker and Posch (2019), for example, we describe our PD work with groups of neurodiverse children to design digital technology for social play. The process specifically aimed to work with controversies as a resource for design. We have argued that the concept of agonistic design has allowed us to resist the temptation to solve every conflict that arose in the design process but rather turn it into a quality of the outcome. All of the resulting designs created agonistic spaces for negotiating intra-action that nurtured constructive conflict in use-time. Similarly, Carl DiSalvo (2012) provides many examples of how what he calls adverserial design creates spaces of conflict in use that allow us to engage in a discourse about power and agency.

So, how can these concepts of infrastructuring, meta-design and agonism help to configure participation around the design-use of smart everyday things? To exemplify the abstract discussion above, I return to the running case of a fitness tracker and sketch a speculative, alternative future:

> SmarterThings, a technology company in the area of wearable computing, decides to move into the booming self-tracking market with their own fitness tracker. In their market research, they identify the growth in this sector, but also that behind the sales numbers, a much more nuanced picture emerges about their use, abandonment and impact in the real world. They aim to do things differently: the design team engages a number of stakeholders in a participatory design process, including potential users (sports enthusiasts, office workers, people with disabilities or health conditions), public health policy makers, medical professionals, privacy lawyers etc. In the design process, stakeholders, designers and technologists negotiate potential meanings of digital technology that can sense body activity. This negotiation is materially discursive, i.e. mainly

facilitated through the collaborative making and trying out of prototypes that foreground the technology as a non-neutral actor itself.

Many tensions, dilemmas and controversies surface in the design process, e.g. around the inbuilt normativity of bodies. But instead of succumbing to trade-offs and compromises, the design team feeds off these controversies and works to create a smart Thing that allows future users to work through these controversies themselves and build their own relationship with it. Their product, SmarterFit, uses AI to be able to evolve its own program of action. Its physical design is modular and open, with many components intended to be tinkered with. SmarterThings infrastructures its products with a community portal that provides, traditional functions such as forums, but also open-sourced 3d-printing files and opportunities for (re-)sharing physical designs of the hardware. The portal also provides nuanced data sharing features within chosen communities of practice (e.g. a football team, hobby-cyclists …).

Sue has bought a SmarterFit in one of its standard configurations with a wristband. She wants to keep track of her new exercise routine and hopes to make it stick by scaffolding it with the use of a fitness tracker. She first gives her new companion a name, she calls it BB, after the Star Wars droid BB-8. Sue wears BB for her first session in which it quietly records all data generated by its sensors. Sue revises the information on her smart phone after the workout, without sharing it or comparing it to averages. Neither of the two have yet established what roles they will play in each other's existence and how they will intra-act. Behaviour patterns have not yet been manifested, i.e. it is not yet determined what BB will be and who Sue wants to be in relation to BB. In the beginning it is a negotiation between the initial agendas of 'sustaining exercise' (Sue) and 'getting the wearer to move' (BB), but neither has figured out yet what that means in their everyday relationship.

After months, their intra-actions have changed. Sue still aims to sustain her exercise, but has realized that family life, work and other circumstances put additional constraints on her routine. For some aspects, she needs more flexibility, particularly in spreading out the time and duration of the exercise. Some data has become critical for her to better manage her Diabetes, e.g. by estimating blood sugar levels in the course of physical activity. BB has changed too, both in its materiality and its program of action. There is the basic sensor module that provides routine activity levels to Sue's blood sugar tracking app. Sue has long had an interest in crafts and now created a whole range of jewelry which all can incorporate BB's sensor module in unobtrusive ways. BB's AI has learnt to not do anything, but blend into the background and provide key data points to the blood sugar app. But there is also the other BB, an upper arm band that Sue created at a local fablab. It combines the full range of sensors and a speech and tactile interface. Whenever Sue wears this arrangement, she

> intra-acts with a much more present BB. She accepts BB's role as her fitness trainer and BB's AI has learnt the motivational strategies that work for Sue. However, BB also routinely challenges Sue's optimization goals, providing opportunities for BB to learn and for Sue to critically reflect. Not all agendas are negotiable to the same degree. For example, BB will never support Sue in clearly unhealthy behaviours, such as not moving at all or a worrying decline of the body-mass-index, and Sue is not willing to be exposed to social peer pressure. Their relationship may still come to an end if any of these red lines are crossed.
>
> Sue and BB also have developed a practice around using SmarterThings' online platform. Sue realized over the first few months with BB just how much of her life can be known through what BB senses and she felt a clear need to protect her privacy. BB probed her stance on privacy at various occasions and adapted accordingly. But Sue also saw the value of having real-world data available to discuss with her doctor and BB helped Sue to set up a secure way to share the right level of information. Sue also joined a running club that had a less competitive, but more supportive approach to its community of practice. Again, BB supported Sue in finding the right level of information that she feels comfortable sharing and which helps sustain the positive vibe in the group. An unexpected outcome of bringing BB into her life was Sue's renewed interest in jewelry making. She now is an active member of a local community with regular meetups at which new ways of blending digital technology such as BB with everyday objects are created at a shared workshop.

The above speculative vignette aims to highlight the shifting roles of people and their smart things and what could happen if spaces for negotiations are designed for. Both Sue and BB in the above story have not known in the beginning where this relationship might be going. Both had initial agendas and goals, but both have evolved in their relationship. Sue has shifted her aims while learning what BB can offer, and BB's program of action has changed along the meaning-making process between the two. The agential cuts that have been agreed on range from a complete embodiment and internalization of a crucial life-support system to an outspoken, external entity that motivates Sue during exercise.

Importantly, such negotiations never end. With constantly changing circumstances, relationships need to be reconfigured and new roles and agential cuts need to be agreed upon. Sue's medical needs may change or SmarterThings may release a SmarterBlood system, changing the device ecology in which BB operates. My central argument here is that this kind of continuous meaning-making is only possible if we design negotiation spaces for these materially discursive practices.

Conclusion

In this chapter, I have sought to use entanglement theories as the theoretical underpinnings to argue for a new conceptualization of the relationship between humans and smart things. Motivated by the increased significance of digital technology in and for our lives, I have harnessed this thinking to argue for new forms of participation across design and use times. Drawing on PD, infrastructuring, meta-design and agonism, I sketch a speculative future in which smart things and us are in a constant state of agonistic negotiation about what we want them and us to be. This requires, I argue, a new smartness and reorients design towards creating objects with intent alongside infrastructures and ecologies that open up spaces for negotiating meaning.

Bibliography

Barad, K. (2007). *Meeting the universe halfway: Quantum physics and the entanglement of matter and meaning.* (2nd print edn). Durham, NC: Duke University Press.

Björgvinsson, E., Ehn, P., & Hillgren, P-A. (2012). Agonistic participatory design: Working with marginalised social movements. *CoDesign, 8*(2), 127–144.

Bonfert, M., Spliethöver, M., Arzaroli, R., Lange, M., Hanci, M., & Porzel, R. (2018). If you ask nicely: A digital assistant rebuking impolite voice commands. In *Proceedings of the 20th ACM International Conference on Multimodal Interaction. ICMI '18* (pp. 95–102). New York: ACM.

Clawson, J., Pater, J. A., Miller, A. D., Mynatt, E. D., & Mamykina, L. (2015). No longer wearing: Investigating the abandonment of personal health-tracking technologies on craigslist. In *Proceedings of the 2015 ACM International Joint Conference on Pervasive and Ubiquitous Computing - UbiComp '15* (pp. 647–658). Osaka, Japan: ACM Press.

DiSalvo, C. (2012). *Adversarial design.* Cambridge, MA: MIT Press.

Ehn, P. (1989). *Work-oriented design of computer artifacts.* (2nd edn). Stockholm: Arbetslivscentrum.

Ehn, P. (2008). Participation in design things. In *Proceedings of the Tenth Anniversary Conference on Participatory Design 2008. PDC '08* (pp. 92–101). Indianapolis: Indiana University Press.

Fischer, G., & Giaccardi, E. (2004). Meta-design: A framework for the future of end-user development. In H. Lieberman, F. Paterno & V. Wulf (Eds), *End user development – empowering people to flexibly employ advanced information and communication technology* (pp. 427–457). Dordrecht, The Netherlands: Kluwer Academic.

Foucault, M., & Gordon, C. (1980). *Power/knowledge: Selected interviews and other writings, 1972–1977.* (1st American edn). New York: Pantheon Books.

Frauenberger, C. (2019). Entanglement HCI The next wave? *ACM Trans. Comput.-Hum. Interact., 27*(1), 1–27.

Frauenberger, C., Foth, M., & Fitzpatrick, G.(2018). On scale, dialectics, and affect: Pathways for proliferating participatory design. In *Proceedings of the 15th Participatory Design Conference: Full Papers – Volume 1* (pp. 1–13). New York: ACM.

Frauenberger, C., Spiel, K., Scheepmaker, L., & Posch, I. (2019). Nurturing constructive disagreement – agonistic design with neurodiverse children. In *Proceedings of the 2019 CHI Conference on Human Factors in Computing Systems. CHI '19* (pp. 1–11). New York: ACM.

Ihde, D. (1990). *Technology and the lifeworld: From garden to earth*. Bloomington: Indiana University Press.

Introna, L. D. (2014). Towards a post-human intra-actional account of sociomaterial agency (and morality). In P. Kroes & P-P. Verbeek (Eds), *The moral status of technical artefacts: Philosophy of engineering and technology* (pp. 31–53). Dordrecht: Springer.

Karasti, H. (2014). Infrastructuring in participatory design. In *Proceedings of the 13th Participatory Design Conference: Research Papers – Volume 1* (pp. 141–150). New York: ACM Press.

Kensing, F., & Blomberg, J. (1998). Participatory design: Issues and concerns. *Computer Supported Cooperative Work (CSCW)*, *7*(3), 167–185.

Kranzberg, M. (July 1986). Technology and History: 'Kranzberg's Laws'. *Technology and Culture*, *27*(3), 544–560. doi:10.2307/3105385. JSTOR 310538.

Latour, B. (2005). *Reassembling the social: An introduction to actor-network-theory*. Clarendon lectures in management studies. Oxford: Oxford University Press.

Ledger, D., & McCaffrey, D. (2014). *Inside wearables: How the science of human behavior change offers the secret to long-term engagement*. Cambridge, MA: Endeavour Partners, LLC.

HIS Markit (2017). *The internet of things: A movement, not a market*. Retrieved 4 March 2021 from https://cdn.ihs.com/www/pdf/IoT_ebook.pdf.

Mouffe, C. (2013). *Agonistics: Thinking the world politically*. London: Verso.

Müller, C. J. (2016). *Prometheanism: technology, digital culture, and human obsolescence*. Critical perspectives on theory, culture and politics. New York: Rowman & Littlefield International.

Muller, M. J. (2003). Participatory design: The third space in HCI. In J. A. Jacko & A. Sears (Eds), *The Human-computer Interaction Handbook* (pp. 1051–1068). Hillsdale, NJ: L. Erlbaum Associates.

Nygaard, K., & Terje Bergo, O. (1975). The trade unions – new users of research. *Personnel Review*, *4*(2), 5–10.

Orlikowski, W. J. (2010). The sociomateriality of organisational life: Considering technology in management research. *Cambridge Journal of Economics*, *34*(1), 125–141.

Rosenberger, R., & Verbeek, P-P. (Eds). (2015). *Postphenomenological investigations: Essays on human–technology relations*. Lanham, MD: Lexington Books.

Rozendaal, M. C., Boon, B., & Kaptelinin, V.(2019). Objects with intent: Designing everyday things as collaborative partners. *ACM Transactions on Computer-Human Interaction*, *26*(4), 1–33.

Spiel, K. (2019). Body-positive computing as a means to counteract normative biases in fitness trackers. *XRDS*, *25*(4), 34–37.

Star, S. L., & Ruhleder, K.(1994). Steps towards an ecology of infrastructure: Complex problems in design and access for large-scale collaborative systems. In *Proceedings of the 1994 ACM Conference on Computer Supported Cooperative Work* (pp. 253–264). New York: ACM.

Stiegler, B. (1998). *Technics and Time, 1: The Fault of Epimetheus*. (Richard Beardsworth & George Collins, Trans.). Stanford, CA: Stanford University Press.

Verbeek, P-P. (2008). Cyborg intentionality: Rethinking the phenomenology of human–technology relations. *Phenomenology and the Cognitive Sciences*, *7*(3), 387–395.

Verbeek, P-P. (2015). Beyond interaction: A short introduction to mediation theory *Interactions*, *22*(3), 26–31.

Woods, H. S. (2018). Asking more of Siri and Alexa: Feminine persona in service of surveillance capitalism. *Critical Studies in Media Communication*, *35*(4), 334–349.

Zuboff, S. (2019). *The age of surveillance capitalism: The fight for a human future at the new frontier of power*. (1st edn). New York: PublicAffairs.

5 THE DYNAMIC AGENCY OF SMART OBJECTS

Jelle van Dijk and Evert van Beek

'Smart objects' are here to stay. Increasingly, there will be wirelessly connected interactive devices and systems in our homes, schools, care facilities and in public spaces, responding intelligently to the world using machine learning (Miorandi, Sicari, De Pellegrini & Clamtac, 2012; Hoffman & Novak, 2015; Alpaydin, 2016; Lemley, Bazrafkan & Corcoan, 2017). At this moment 34 per cent of US households use what is called a *smart assistant* (Petrock, 2019). Smart Internet of Things (IoT) is used in rehabilitation and care homes (Chan, Estève, Escriba & Campo, 2008). In some cities pedestrians now share the pavement with delivery service robots, independently making their way to the next customer (Hawkins, 2019). The South Korean government decided to provide free smart speakers to the elderly, with the aim of reducing loneliness (Volkskrant, 2019). 'Smart IoT' is transforming from a luxury gadget into something that may soon become a basic, ubiquitous aspect of everyday life (Risteska Stojkoska & Trivodaliev, 2017).

The artefacts considered here are often presented as 'smart *agents*': autonomous, assistive and conversational. Terms like *smart* and *agent* are, however, suggestive. They easily conjure up the image of an artificial, thinking *other*, not unlike ourselves. Thus, instead of a mere machine, we take the vacuum cleaning robot to be a new member of the household, roaming the floors with 'a mind of its own'. The appearance and interactions offered by most *social robots* are even designed explicitly with this frame in mind (Coeckelbergh, 2009; Zaga, Lohse, Truong & Evers, 2015; Zaga, 2017). To see smart objects as thinking, social others betrays influence from sci-fi literature and film, but the idea is equally fostered in academic research. Cognitive science has always been concerned with what it would take for a system to be a truly 'intelligent agent'. And in pursuing this question, human intelligence has typically been used as the measure of comparison, the famous

Turing test, which asks whether a computer's responses can be distinguished from those of a human, being the paradigmatic case in point (Turing, 1950). It is important, however, to emphasize that in much of the debate on whether artificial intelligence (AI) is possible, the human we compare machines to is itself already from the outset assumed to be a specific sort of agent, namely, an internal information processing system, implemented in the brain, representing and thereby exerting control over a world 'outside'. That is, not only do we perceive smart devices as crude mirror images of ourselves, but in doing so, we already first interpreted ourselves to be the thinking agent as it is theorized by the dominant Cartesian tradition in cognitive science.

AI and its underlying assumptions have generated much, mostly unresolved, philosophical debate (Turing, 1950; Dreyfus, 1972; Weizenbaum, 1976; Searle, 1980; Haselager, 1997; Hayles, 1999; Kurzweil, 2010; Müller & Bostrom, 2016). Industry, meanwhile, embarked on a more pragmatic journey. Here, terms like 'smart' and 'agent' are used rather loosely. Typically, any device with sensors, actuators and some amount of digital information processing is called smart or intelligent (Rose, 2014; Noessel, 2017). Yet, for all the progress made, no 'smart' device, and certainly not consumer products like 'home assistants' or vacuum cleaners, will pass the Turing test. Although 'smart' is painted in capital letters on the box, most people would readily judge these devices to be rather 'dumb'.

The 'smart agent': A limiting frame

As AI and IoT enter everyday life, smart objects have become a hot topic in interaction design. Many designers are keen on using AI. For interaction designers, the goal is, however, not to design intelligent algorithms as such: The aim is to design certain experiential qualities *in the interaction* between smart objects and their users. Designers know well that the limited technologies they work with are not 'truly' intelligent. In fact, a significant part of the job is to create interfaces that compensate for a *lack* of algorithmic intelligence. Luckily, people only need a minimum of encouragement to read autonomy and smartness into the responses of interactive technologies (Weizenbaum, 1966; Ihde, 1990; Turkle, 2010). Interaction design employs a variety of 'tricks' to help suspend disbelief, with the hope to keep up appearances long enough for the interaction to unfold seamlessly, relative to the task at hand.

Unfortunately, in practice, believable simulations do tend to break down all too quickly. Present-day smart objects misinterpret events or user requests that even a child would understand, make errors that can put people in serious danger, give rise to the 'uncanny valley' experience and so on. Examples are the automated service phone calls of Google Duplex (O'Leary, 2019), navigation systems not incorporating context and sending drivers astray (Lin, Kuehl, Schöning & Hecht,

2017), Amazon's Alexa ordering people dollhouses after hearing its name on TV (Liptak, 2019), a patrolling security robot knocking down a toddler (Vincent, 2016) and so on. As users are, for better or for worse, adapting their communication styles to match the demands of these poorly performing machines, our appreciation for the quality of real human social interaction may erode. Sherry Turkle (2015) worries that AI that is designed to mimic human interaction, yet fails in the attempt, may 'drag us down' as it were, reducing the quality of human life rather than enriching it.

At this point, one response could be to simply try and 'do better'. AI engineers could work harder on creating even smarter algorithms, and interaction designers could become more sophisticated in creating suggestive interfaces to smooth out any remaining wrinkles. Such efforts would improve the illusion that the user is interacting with an intelligent 'other' agent. In this chapter, however, we pursue a different direction, which starts by first questioning this dominant frame of the smart agent itself.

Smart agents: What frames do we have?

The 'thinking other agent' is just one lens through which to look at smart technologies, one that in practice quickly leads to disappointment. As long as interaction designers consider it their main task to 'cover up' for the lack of actual intelligence in 'smart' algorithms, design research remains limited in thinking more fundamentally about possible roles 'smart' objects could take. Both industry as well as traditional philosophical debate in AI have largely overlooked how artefacts currently sold under the name of 'smart' are actually experienced in real life (Suchman, 2006; Zaga *et al.*, 2015; Kudina, 2018). If we look at how smart objects are actually experienced, the question is not whether the object is smart, or appears so. The question becomes: What does this object *do* (Verbeek, 2000)? That is, how does this concrete technological totality of sensing, actuation, machine learning and physical design impact the user's experience in daily life? Rather than already assuming what a smart object is supposed to be, demanding that it lives up to the expectation of being a Cartesian agent 'with a mind of its own' – we take a step back and consider the diversity of possible forms of agency artefacts may adopt once they become elements in human practices. We sketch out how to design with openness towards such emerging roles of smart objects that will occupy our lives.

An embodied perspective

To be able to discuss smart objects in the context of their participation in concrete, everyday human practices, we adopt an embodied perspective on human-technology interaction (van Dijk, 2018). This perspective draws from theories of embodied and enactive cognition (Clark, 2003; Di Paolo, Cuffari & De Jaegher, 2018) as well as from the work of Maurice Merleau-Ponty, whose phenomenology centres on our embodied 'being-in-the-world' (Merleau-Ponty, 1962). Embodied theories explain how human beings make sense of the world through engaging in immediate interaction with the local environment. Importantly, this environment includes designed artefacts and technologies. Embodied theories generally present a bottom up, self-organizing (Kelso, 1995), action-oriented (Clark, 1998) and anti-Cartesian, 'in the world' perspective on human action and sense-making (Heidegger, 1927; Merleau-Ponty, 1962; Dreyfus, 1972, 2002). On the embodied view, intelligent action emerges from the living moving body, situated in its environment, and not from an internal, thinking 'mind' (De Jaegher & Di Paolo, 2007). Furthermore, our lived (i.e. 'experiential') bodies are not just physical objects in space: We are social bodies situated within human practices at the same time (Suchman, 1987). The experiential, material and social world are often dissociated in analysis, but embodied theory considers this to be misleading. For a human sense-maker, the *lifeworld* presents itself as one integrated meaningful whole, with a material aspect as well as social, cognitive and emotional significance, all tightly interwoven and co-determining (Schutz & Luckmann, 1995; Agre & Horswill, 1997; Di Paolo *et al.*, 2018).

Being situated in the lifeworld, human beings engage in ongoing activity towards attaining 'grip' on the situation at hand (Dreyfus, 2002; Merleau-Ponty, 1962). Through interacting, we become 'attuned' to our lifeworld. Over time, initially volatile couplings between perception and action may develop into more stable skills and habits. With each new skill, the lifeworld opens up in new ways (Merleau-Ponty, 1962). Consider how a professional painter 'sees' quite a different object than does the unskilled homeowner, even as both are together standing in front of the same wall. This is what enactivists mean when they say we 'enact' our world (Varela, Thompson & Rosch, 1991). As said, artefacts, such as tools, figure in these world-enactments, forming a binding anchor point around which the interaction unfolds (Merleau-Ponty, 1962). When a skilled person picks up a tool, this generates a qualitative change in perception, and with it a new field of affordances generated by having that tool in (a skilled) hand (Gibson, 1979; Heidegger, 1927; Rietveld, De Haan & Denys, 2012). In this regard, Merleau-Ponty discusses the incorporation of tools, such as the blind man's cane, into our lived body (Merleau-Ponty, 1962). Likewise, Andy Clark discusses how external technologies may become cognitive extensions (Clark, 2003). Technologies and the skills that evolve with them are part of sociocultural practices. Our lifeworlds are not just social in nature because they literally contain other humans: They are filled with artefacts and designed spaces created by, and in their operation further sustaining, cultural practices and social life (Lave, 1988; Hutchins, 1995; Rietveld *et al.*, 2012).

With this framework of humans as embodied sense-makers in place, we now question the position of smart objects. Bracketing the frame of the autonomous, thinking agent, we see smart objects first as *mediating elements* in embodied practices. The question is not how smart objects can provide a believable simulation; the question is how they enact various forms of agency in practice whilst being appropriated within the couplings between humans, other artefacts and the wider lifeworld. To explore these forms in a structured way, we introduce mediation theory (Verbeek, 2015). Adding to the embodied perspective, mediation theory allows describing various ways in which smart objects *mediate* our relation to the world.

Mediation theory and smart agents as 'others'

Peter-Paul Verbeek (2015), building on Don Ihde (1990), proposes various human-world relations, all of which are mediated by the tools and technologies that form a part of our everyday life. In this picture, agency is not a pregiven fact of the matter or a property of human or artefact. Their post-phenomenological view rejects the idea that humans and technologies first have a self-contained, predefined form of agency, upon which the two then inter-act. Agency emerges as a quality existing in the interaction between people and artefacts. In line with the embodied perspective outlined earlier, Verbeek sees relations with artefacts as part of a larger relation between humans and their world more broadly, which is mediated by technology, and of which technologies are a part. 'What is being designed, then, is not a thing but a human-world relation in which practices and experiences take shape' (Verbeek, 2015; see also Suchman, 1987). Ihde initially describes four types of relations, ranging from technologies as fully embodied to technologies fully in the background of awareness.

The embodiment relation describes a relation where a technological artefact is fully incorporated in ones' bodily comportment towards the world. This means that one's focus is directed, rather than at the artefact, at the world that is 'beyond the artefact'. An important quality of these technologies is their transparency. Examples given are the phone through which we speak with other people, rather than with or to the phone itself. Glasses mediate our experience of the world, but once we get used to them we do not notice their presence: they have become part of our lived bodies with which we perceive the world. Ihde presents a schematized representation of this relationship as: (human-technology) → world. The technological artefact forms a unity with the human being.

Hermeneutic relations are relations in which technology does not offer access to the world transparently, like glasses, but via technological instruments that represent aspects of the world. This representation needs to be interpreted: Our engagement with the world is hermeneutic. We make sense of the world by reading a representation of the world presented on a medical scan on a monitor. The beep of a metal detector represents a piece of buried metal, the height of fluid in the thermometer represents the temperature. In this case, technologies form a unity with the world: Human → (technology-world).

Alterity relations describe interactions with technologies where the world fades largely in the background. The human being is engaged with the technology rather than with the world around this technology. The role of the artefact becomes that of a 'quasi-other'. Examples of these interactions include the interaction one engages in with the ATM to get money out of it ('Would you like to make a withdrawal?'), or the 'dialog box' that opens on a computer screen to provide program installation instructions. Even more prominent as a 'quasi-other' is the social robot in human-robot interactions. The alterity relation comes closest to the common-sense view of smart objects as thinking, social agents users would interact with as they would with other human beings. However, mediation theory does not literally mistake such devices to be actual 'people'; it simply describes how these interface modes take an analogous form. In schema this relation can be represented as: human → technology (world).

The fourth relation is the background relation. Here technology becomes an element of the implicit, taken-for-granted background, which operates outside of our awareness. Although we do not actively engage with them, background technologies provide important context for our sense-making. Think of the light in the room, or the central heating system providing basic comfort. Their implicit supportive role in our practices usually becomes visible only when they break down. This relation is schematized as: human (technology/world).

Many recent technologies, however, do not fit neatly into one of these four types. Verbeek (2008) indicates how the intentionality of human and technology can blend together in a hybrid intentionality. A new entity comes about in a cyborg relation. Think of someone with a brain implant where human and technology are physically merged. Other technologies merge with our environment. Similarly, an embodiment relation offered by 'augmented glasses' breaks down in the instant a message appears in one's field of vision. Persuasive, ubiquitous, tangible technologies and ambient intelligence are actively changing the lifeworld, and they ask for new understandings in terms of mediation. It is an open question whether smart objects would, and should, necessarily give rise to an alterity relation. So, let us take an empirical approach. In the next section we present two case studies. These studies are discussed from the perspective of the design-researcher, meaning that we are concerned with the decisions on practical details in an iterative design process, and we analyse what they would mean for the quality of resulting interaction. With these concrete design outcomes in mind we discuss the various mediating roles that smart objects may play.

Design cases

BagSight

BagSight is a leather backpack that can move itself on the back of the wearer and react to environmental stimuli (Figure 5.1). Developed as a research-through-design artefact, it helped to investigate ways in which people experience a relationship with a 'smart', interactive object. BagSight can be seen as an embodied artefact that has some 'autonomy' in the sense that it responds to what it senses and displays expressive behaviour.

The study was part of a larger project, 'Objects with intent', in which indeed smart objects are framed as autonomous agents, collaborative partners with a focus on intentions (Rozendaal, Boon & Kaptelinin, 2019), In this particular study, metaphors associated with 'the intelligent other' were used as a design tool, especially building on the metaphors introduced by Braitenberg's Vehicles (Braitenberg, 2004). The Vehicles are a series of elementary robot designs easily perceived by a human observer as intentional. Initially design moved towards an alterity relation, imagining an object as an intelligent other and imagining the backpack as a technological analogue to the guide dog. A first iteration conceived of the object as a divining rod, held by hand. Several iterations on the design led to the insight that changing the location of the object to the back of the user could lead to a more multifaceted interaction. Among these facets are the intimate qualities of something placed around the shoulders and the everyday character of a backpack.

These developments finally led to BagSight, exploring how the backpack could take on the role of a travel companion while also being an everyday use artefact. Inspired by the Vehicles, the backpack is designed to exhibit two types of behaviour that blend while wearing. First, BagSight's behaviour is designed to be 'afraid' of obstacles, meaning that it moves to the left on the back of the wearer when the distance sensor observes an obstacle on the right. The second type of behaviour

FIGURE 5.1 BagSight moves on the back of the wearer and is 'afraid of obstacles'.

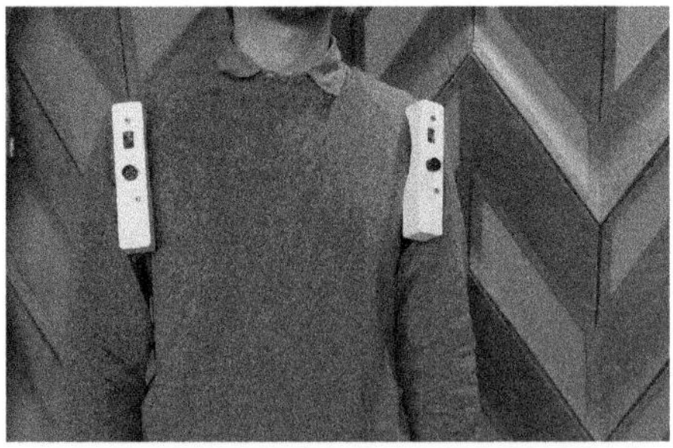

FIGURE 5.2 Two distance sensors and two light sensors provide the input for BagSight.

is an attraction or 'love' of light. It moves to the right when the light sensor on the right shoulder pad observes more light. This behaviour is realized through an Arduino microcontroller that controls two servo motors rolling up the left or right cord of the backpack, effectively moving BagSight left and right and up and down on the back of the wearer in an area of about 30 cm wide and 40 cm high. The input comes from two forward and slightly outward-facing distance sensors and two light sensors mounted on two modules on the front of the shoulder, around 10 cm below shoulder level (Figure 5.2).

Evaluation of BagSight

BagSight has been employed in an experimental study with sixteen participants. Before the experiment, BagSight was introduced to the participants as 'a smart backpack'. Participants are asked to navigate the environment using the backpack. Afterwards participants were asked 'what the backpack wants'. These interviews were audio recorded; relevant quotes were transcribed. A shared phenomenological reduction was applied and the quotes are structured in themes in two iterations by two researchers.

How people experience BagSight

When participants employed the backpack to navigate an environment, they described their interactions with this smart object and the environment in different ways, which we can relate to the various forms of mediation as introduced in post-phenomenology.

Some comments described the interaction as 'perceiving the environment through an incorporated object'. Participants described their perception of the environment without directly relating this to the behaviour of the backpack. They used it as an extension of their sensory capabilities, for example, 'sort of like your eyes.' Here we see an embodiment relation: 'I see it as more and more a unity with myself, as I learned to walk with it.' And 'I see it as an extension of the human senses'.

We also find the interaction described as a 'perception of the environment through an independent observer'. Here participants described how they experienced the changes in the artefact. These quotes emphasized how BagSight functioned as a measurement instrument that needed to be 'read' to estimate distances and to move around. 'At first I did not quite understand how the backpack worked. At a certain moment you start to understand, I need to interpret this signal from the backpack like this. If it does this, I have to continue straight ahead.' Interpreting the feedback allowed them to make sense of what was going on. BagSight was described as having its own interaction loop that the wearer was observing hermeneutically.

Some quotes hint at experiencing interactions with 'another', in a social sense, either a 'leader', instructing them where to go, or a 'buddy', a companion that made the situation more like going on an exploration together and helping each other out in finding one's' way: 'He was the navigator.' They used anthropomorphic descriptions emphasizing how they were guided around and following commands of the object. 'It warns you, by getting loose or tight when you approach an object.' This stance is similar to a popular way of describing the interaction with a GPS navigation device. 'It wanted me to go in this direction.' This alterity relation ranged from a pure 'leader' interpretation to 'perceiving being accompanied by a buddy'. For example, a comment like 'It tries to prevent you from walking into a wall' emphasized companionship, with backpack and wearer shaping each other's behaviour. 'It wanted to go to that location.' While anthropomorphic, such comments were more similar to relationships in guide dog teams, than simply 'being told what to do'. Participants also made clear how a more subtle interaction (gentle pushing) distinguished this backpack from their image of a 'machine'.

Reflections on BagSight

BagSight explored ways in which an artefact can take on some form of agency, as perceived by its user, when used in a concrete situation. While the initial goal and the framing of the experiment ('What did the backpack want?') portrayed BagSight as a social other, it turned out in the evaluations that alterity is just one type of relation that people may have with BagSight. Furthermore, dividing lines between types of relations are blurred and dynamic, and individual people were not consistently perceiving the backpack in just one way or another.

BagSight can be 'a companion that walks with you' but also 'an extra pair of eyes', or 'an augmented way of vision', and these two modes of 'being' of the artefact can flow back and forth. Verbeek and Ihde call this multistability: one artefact may install several sorts of mediation relations, and a person's experience may switch between these types similar to the way our perception of the Necker cube switches at intervals.

At the same time, BagSight provides an interesting case for comparison with the cane for visually impaired, as, for example, described by Merleau-Ponty (1962). The cane, as a visible signal of impairment, scaffolds a social interaction between humans, for example, for giving way in traffic and in some cases even leading to an aversion of social interaction from bystanders. As a less conspicuous artefact, BagSight removes this scaffold in part. However, when its movements become visible to onlookers, it may form a stepping stone for a different type of social interaction, one of interest and wonder. This social scaffolding, well described in embodied theories such as in De Jaegher and Di Paolo (2007) and Suchman (1987), has no clear mapping onto the relations in mediation theory.

The question is to what extent a designer can influence the emergence of one mediation relation or another. We think that, given that the BagSight project explicitly tried to create an alterity relation, it is crucial to the mixed results that the form and behaviour of the backpack did *not* resemble the traditional humanoid shape of the social other. The backpack could not move around in space on its own, it did not have its own 'head' that could attend to the user in a social manner and the user could not attend to the backpack in the social manner custom between human beings (or even animals). The backpack had particular features that contributed to the embodiment experience. For example, it is worn on the body, but more importantly, its sensors and actuators are being designed such as to 'perceive along with' the user in an ego-centric fashion, which make it more likely for the user to perceive the artefact as a sensory 'augmentation'. At the same time, the study suggests that designers can never fully determine what composition of agency will emerge; there is always multistability.

In this case study we explored a number of different forms of agency that an interactive artefact may take on, from being a 'social other' that directs, to one that 'co-shapes', to an instrument that provides information that needs to be interpreted, to an embodied augmentation of the user's own agency. In the next section we look at these same relations, yet focusing on the user's perspective: How does the user's agency change as mediated by the form and behaviour of interactive artefacts?

Highlight

This case study involves a four-year participatory, research-through-design project (Stappers & Giaccardi, 2017) grounded in embodied phenomenology and embodied cognition theory. The objective was to enrich the daily living environment of young adults on the autistic spectrum and to empower these young adults in living a (more) independent life. The concrete design project, in which next to designers' autistic adults, family members and care professionals

FIGURE 5.3 Highlight: Wireless luminous objects help organize and execute daily activities.

participated, served to ground theoretical reflections about the distribution of agency and control in the management of everyday life, between the autistic user, important others and technology. The discussion that follows resulted from an iterative process of structured reflection on (participatory) design action as well as contextual evaluations over a period of four years, involving seven young adults on the autistic spectrum and their care professionals, using intermediate prototypes as scaffolds for creating shared understanding.

The Highlight concept

The project focused on managing everyday activities in the home. The design that resulted, called Highlight, is a system of wireless luminous objects that literally highlight parts of the familiar structure of things in the apartment (Figure 5.3). The wireless objects light up in a location in the apartment at the time a planned activity was scheduled to take place at that location (Figure 5.4). They put focus on the current task at hand and invite to move on to the next activity when the time is ripe. The assumption is that Highlighting the relevant place at the relevant time produces the correct action affordance. This choice came out of feedback from participants themselves, who typically have good memories but may have trouble in getting from intention to action and switch tasks. Also, participants indicated the wish to be in control: The system should not give instructions but rather a subtle invitation that a person is still free to respond to.

Reflections on Highlight

While BagSight explored agency from the perspective of the artefact in relation to user and context, the Highlight project can be seen as having the opposite starting point. Here, the aim was to 'empower', that is, to increase the agency of the user and investigate how supportive 'smart' technologies in the local environment could contribute to such agency.

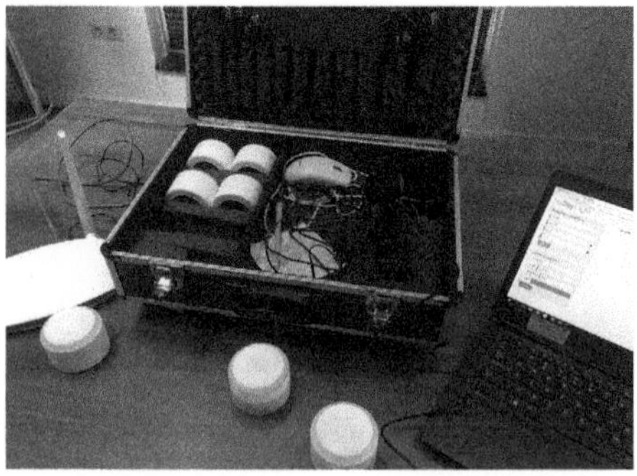

FIGURE 5.4 Fully working prototype kit of seven lamps and interface, as used in evaluations.

FIGURE 5.5 Highlight in use in a facilitated living apartment by a young autistic man.

Early in the project there was only one lamp, designed to *instruct* the user. It effectively told the user what to do and when. Like the 'leader' role of BagSight, this suggests an 'alterity' relation: One looks actively at the lamp and the lamp provides an explicit message. At the same time, however, the lamp also indicates a hermeneutic relation: It was a stand in for a caregiver who is not physically present to give the instruction. However, our autistic participants wanted to be supported, but not told what to do when. They wished to remain in control. In co-designing the lamps, the autistic participants and their care workers together negotiated a form of assistance that would help the person where needed, without 'taking over control'. The lamps took more the role of being a 'companion', as we also saw in BagSight. Meanwhile, instead of one central lamp, the system developed into an ambient network of luminous elements, which moved towards a *background* relation. This did not mean the lamps disappeared completely outside of awareness. Highlights are present, albeit in a subtle, peripheral way (Bakker, van den Hoven & Eggen, 2013). The lamps 'highlight' certain areas of the living room, which would draw a person's attention to that area, and this would influence a person's affordances for action (Figure 5.5). In embodied cognition theory we could say the lamps play a mediating role in sustaining sensorimotor couplings: Changing the distribution of salience of stimuli, they guide attention and action. This would correspond to an embodiment relation. Finally, the concept included various ways for users to freely explore their own desired routines, by changing the location and timing of the lamps themselves, and learning from its effects. In this scenario the lamps act more like 'notes to self'. This could somewhat paradoxically be seen as once again an 'alterity relation with oneself': You tell yourself what to do and learn by doing.

In summary, we may say that the lamps provide for a complex, hybrid mediated experience: At the moment the lamps help to sustain a habitual action-perception routine, they can said to be embodied as well as partly background, yet when a person is actively organizing and reorganizing the lamps they take on a hermeneutic as well as an alterity role: a reflective conversation with oneself.

Discussion

Present-day smart objects such as home assistants like Alexa, Google or Siri and various social robots remain firmly rooted in traditional cognitive science conceptions. Both user and smart object are considered to be self-contained 'thinking agents'. In mediation theory, interacting with such agents boils down to an alterity relation: engaging with a 'quasi-other'. Given that 'true' AI is not yet a reality, interaction designers create, in pragmatic fashion, 'believable simulations' of the 'thinking other agent'. Even though the alterity relation is only one of various possible relations, Rosenberger & Verbeek (2015) readily assume that 'AI' as a technology will generates an ever-increasing dominance of 'alterity' mediation in society:

> Becoming conscious of the various relations that interactive technologies may support, and looking carefully and with an open mind to what devices and systems actually do in real-world human practices, designers can instead use smart technologies to design for a *diversity* of mediating relations.

We argue this is not because 'smart technologies' necessarily produce alterity relations. Rather, alterity is installed by a design grounded in theoretical assumptions about what 'smartness' means, as entertained within the AI tradition and in popular culture. Such assumptions are at risk of being uncritically carried over into interaction design. By reflecting on empirical cases, we deconstructed the idea of the 'smart object' as something already understood, namely, as a Cartesian, thinking other.

Becoming conscious of the various relations that interactive technologies may support, and looking carefully and with an open mind to what devices and systems actually do in real-world human practices, designers can instead use smart technologies to design for a *diversity* of mediating relations. BagSight and Highlight do not resemble anything like a humanoid, 'person', not a 'someone' we can talk to. Both assume human sense-makers situated in lifeworlds and both mediate the moment-to-moment interactions that give rise to a person's immediate lived experience. These interactions turned out to display all forms of agency, depending on context and flow, dynamically transforming, and sometimes entangled in ways that can no longer be separated. Agency is not an a priori given, we see dynamic agency.

Designing for dynamic agency

To open the design space for dynamic agency, we suggest directing our efforts first outwards, designing for interaction, and only later investigate how internal processing may add value. That is, we should design 'where the action is' (Dourish, 2004). Grounded in internalist conceptions of intelligence, machine-learning algorithms are typically designed to be 'autonomously smart', that is, to understand the world and solve a task at hand by internal processing. When designing for dynamic agency, one first designs external form and behaviour, and explores in detail how these forms will become appropriated in practices. Such outward forms are *not* merely a compensation for a lack of intelligent processes inside. It is the other way around: It becomes an open question as to what, for example, machine learning *may add to* an interaction design (cf. Wiberg, 2018); internal 'smartness' may be added as a last resort. In a shift of figure and background, the interface becomes the product and machine learning the 'add-on'. Thus, we ask: What value could more advanced algorithms add to BagSight's current, Braitenberg-style mappings? And would more sensors and processing add value to the guidance currently offered by Highlight?

The role of the designer

How to approach this newly opened design space? The designer has to adopt a less authoritative position. The smart object becomes a thing in itself that is open to be appropriated by humans in sense-making practices. Highlight, designed for a hermeneutic relation, was sometimes related to as a stand-in for a caregiver. BagSight, designed to be a companion, was also appropriated as a sensory augmentation. The relation we see in practice is not predetermined by the 'smartness' of the technology. The way the user relates to the object, and through the object, to the lifeworld emerges in different contexts in different ways, and only once it is taken up in a human sense-making process. The question then is how interaction designers can design 'things', if they are not the (only) ones determining mediating relations and configurations of agency. It is the designer who gives form to the thing, a configuration of technology, form and behaviour. These 'things' enable different relations, they are 'multistable', in the terminology of post-phenomenology (Ihde, 2012). In response to this multistability we suggest design strategies that emphasize openness and ambiguity. Several design strategies have been proposed that give form to things which are open to different mediating relations. For example, Sengers and Gaver (2006) and Boon, Rozendaal and Stappers (2018) propose ambiguity as a resource for designers. The end products of a design process could afford multiple interpretations and multiple courses of action. Nicenboim, Kitazaki, Kihara, Marin and Havranek (2018) similarly propose strategies that make the designer aware of human resourcefulness that complements the openness of the things. Seok, Woo and Lim (2014) describe *non-finito* products as being intentionally unfinished, leaving room for creativity of end users in solving their problems. All of these strategies allow for shifts in emerging relationships. They acknowledge a mutual constitution of thing and human not decided solely by designer or technology.

Conclusion

The present analysis pushes back against the alterity relation as the obvious and only possible candidate for things we call 'smart'. The concept of dynamic agency may help designers create novel, interactive devices which users may incorporate into their practices, allowing for new ways of sense-making. In other words, we seek technologies that allow for 'making a difference in the world' (Barad, 1996) rather than creating in machines a *mirror image* of that which we already believe to know about ourselves (Rosenberger & Verbeek, 2015). Dynamic agency therefore calls for smart objects to become productive elements within the ongoing sense-making practices of humans in their lifeworlds.

Bibliography

Agre, P., & Horswill, I. (1997). Lifeworld analysis. *Journal of Artificial Intelligence Research, 6*, 111–145.

Alpaydin, E. (2016). *Introduction to machine learning: Selected papers of Lionel W. McKenzie*. Cambridge, MA: MIT Press.

Bakker, S., van den Hoven, E., & Eggen, B. (2013). FireFlies: Physical peripheral interaction design for the everyday routine of primary school teachers. In *Proceedings of the 7th International Conference on Tangible, Embedded and Embodied Interaction*, 57–64. https://doi.org/10.1145/2460625.2460634.

Barad, K. (1996). Meeting the universe halfway: Realism and social constructivism without contradiction. In L. H. Nelson & J. Nelson (Eds), *Feminism, science, and the philosophy of science* (pp. 161–194). Dodrecht: Springer.

Boon, B., Rozendaal, M. C., & Stappers, P. J. (2018). Ambiguity and open-endedness in behavioural design. In *Proceedings of the DRS 2018 International Conference: Catalyst* (pp. 2075–2085). https://doi.org/10.21606/drs.2018.452.

Braitenberg, V. (2004). *Vehicles: Experiments in synthetic psychology*. Cambridge, MA: MIT Press.

Chan, M., Estève, D., Escriba, C., & Campo, E. (2008). A review of smart homes—present state and future challenges. *Computer Methods and Programs in Biomedicine, 91*(1), 55–81.

Clark, A. (1998). *Being there: Putting brain, body, and world together again*. Cambridge, MA: MIT Press.

Clark, A. (2003). *Natural-born cyborgs: Minds, technologies, and the future of human intelligence*. New York: Oxford University Press.

Coeckelbergh, M. (2009). Personal robots, appearance, and human good: A methodological reflection on roboethics. *International Journal of Social Robotics, 1*(3), 217–221.

De Jaegher, H., & Di Paolo, E. A. (2007). Participatory sense-making: An enactive approach to social cognition. *Phenomenology and the Cognitive Sciences*, 6(4), 485–507.

Di Paolo, E. A., Cuffari, E. C., & De Jaegher, H. (2018). *Linguistic bodies: The continuity between life and language*. Cambridge, MA: MIT Press.

Dourish, P. (2004). *Where the action is: The foundations of embodied interaction*. Cambridge, MA: MIT Press.

Dreyfus, H. (1972). *What computers can't do*. Cambridge, MA: MIT Press.

Dreyfus, H. (2002). Intelligence without representation: Merleau-Ponty's critique of mental representation. *Phenomenology and the Cognitive Sciences, 1*, 367–383.

Gibson, J. J. (1979). *The ecological approach to visual perception*. Boston: Houghton Mifflin.

Haselager, W. F. G. (1997). *Cognitive science and folk psychology: The right frame of mind*. London: Sage.

Hawkins, A. (2019, 20 August). Thousands of autonomous delivery robots are about to descend on US college campuses. *The Verge*. Retrieved 1 February 2021 from https://www.theverge.com/2019/8/20/20812184/starship-delivery-robot-expansion-college-campus.

Hayles, N. K. (1999). *How we became posthuman: Virtual bodies in cybernetics, literature and informatics*. Chicago: University of Chicago Press.

Heidegger, M. (1927). *Sein und Zeit*. Tübingen: Max Niemeyer Verlag.

Hoffman, D. L., & Novak, T. P. (2015). Emergent experience and the connected consumer in the smart home assemblage and the internet of things. *SSRN Electronic Journal*. https://doi.org/10.2139/ssrn.2648786.

Hutchins, E. (1995). *Cognition in the wild*, Cambridge, MA: MIT Press.

Ihde, D. (1990). *Technology and the lifeworld: From garden to earth*. Bloomington: Indiana University Press.

Ihde, D. (2012). *Experimental phenomenology: Multistabilities*. Albany: SUNY Press.

Kelso, J. A. S. (1995). *Dynamic patterns: The self-organization of brain and behavior*. Cambridge, MA: MIT Press.

Kudina, O. (2018), The technological mediation of morality: Value dynamism, and the complex interaction between ethics and technology. PhD thesis, University of Twente, Enschede.

Kurzweil, R. (2010). *The singularity is near: When humans transcend biology*. New York: Penguin.

Lave, J. (1988). *Cognition in practice: Mind, mathematics and culture in everyday life*. Cambridge: Cambridge University Press.

Lemley, J., Bazrafkan, S., & Corcoran, P. (2017). Deep learning for consumer devices and services: Pushing the limits for machine learning, artificial intelligence, and computer vision. *IEEE Consumer Electronics Magazine*, 6(2), 48–56.

Lin, A. Y., Kuehl, K., Schöning, J., & Hecht, B. (2017). Understanding 'death by GPS': A systematic study of catastrophic incidents associated with personal navigation technologies. In *Proceedings of the 2017 CHI Conference on Human Factors in Computing Systems* (pp. 1154–1166). https://doi.org/10.1145/3025453.3025737.

Liptak, A. (2017). Amazon's Alexa started ordering people dollhouses after hearing its name on TV. *The Verge*. Retrieved 1 February 2021 from https://www.theverge.com/2017/1/7/14200210/amazon-alexa-tech-news-anchor-order-dollhouse.

Merleau-Ponty, M. (1962). *Phenomenology of perception*. (Colin Smith, Trans.). London: Routledge.

Miorandi, D., Sicari, S., De Pellegrini, F., & Chlamtac, I. (2012). Internet of things: Vision, applications and research challenges. *Ad Hoc Networks, 10*(7), 1497–1516.

Müller, V. C., & Bostrom, N. (2016). Future progress in artificial intelligence: A survey of expert opinion. In V. C. Müller (Ed.), *Fundamental issues of artificial intelligence* (pp. 555–572). Cham: Springer.

Noessel, C. (2017). *Designing agentive technology: AI that works for people*. New York: Rosenfeld Media LLC.

Nicenboim, I., Kitazaki, M., Kihara, T., Marin, A. T., & Havranek, M. (2018). Connected resources: A novel approach in designing technologies for older people. In *Conference on Human Factors in Computing Systems - Proceedings, 2018-April* (pp. 1–4). https://doi.org/10.1145/3170427.3186527.

O'Leary, D. E. (2019). GOOGLE'S Duplex: Pretending to be human. *Intelligent Systems in Accounting, Finance and Management, 26*(1), 46–53.

Petrock, V. (2019, 20 July). US voice assistant users 2019. *Emarketer*. Retrieved 1 February 2021 from https://www.emarketer.com/content/us-voice-assistant-users-2019.

Rietveld, E., De Haan, S., & Denys, D. (2012). Social affordances in context: What is it that we are bodily responsive to? *Behavioral and Brain Sciences, 36*(4), 436–436.

Risteska Stojkoska, B. L., & Trivodaliev, K. V. (2017). A review of internet of things for smart home: Challenges and solutions. *Journal of Cleaner Production, 140*, 1454–1464.

Rose, D. (2014). *Enchanted objects: Design, human desire, and the internet of things*. New York: Scribner.

Rosenberger, R., & Verbeek, P.-P. (2015), *Postphenomenological investigations: Essays on human-technology relations*. Lanham, MD: Lexington Books.

Rozendaal, M. C., Boon, B., & Kaptelinin, V. (2019). Objects with intent: Designing everyday things as collaborative partners. *ACM Transactions on Computer-Human Interaction, 26*(4), 1–33.

Schutz, A., & Luckmann, T. (1995). *The structures of the life-world*. Evanston, IL: Northwestern University Press.

Searle, J. R. (1980). Minds, brains, and programs. *Behavioral and Brain Sciences, 3*(3), 417–424.

Sengers, P., & Gaver, B. (2006, June). Staying open to interpretation: Engaging multiple meanings in design and evaluation. In *Proceedings of the 6th Conference on Designing Interactive Systems* (pp. 99–108). New York: ACM.

Seok, J., Woo, J., & Lim, Y. (2014). Non-finito products. In *Proceedings of the SIGCHI Conference on Human Factors in Computing Systems* (pp. 693–702). https://doi.org/10.1145/2556288.2557222.

Stappers, P., & Giaccardi, E. (2017). Research through design. In: M. Soegaard & R. Friis-Dam (Eds), *The encyclopedia of human-computer interaction*. Retrieved 1 February 2021 from http://www.interaction-design.org/literature/book/the-encyclopedia-of-human-computer-interaction-2nd-ed/research-through-design.

Suchman, L. (1987). *Plans and situated actions: The problem of human-machine communication.* Cambridge: Cambridge University Press.

Suchman, L. (2006). Reconfiguring human-robot relations. In *ROMAN 2006 – The 15th IEEE International Symposium on Robot and Human Interactive Communication* (pp. 652–654). https://doi.org/10.1109/ROMAN.2006.314474.

Turing, A. M. (1950). I.—Computing machinery and intelligence. *Mind, 59*(236), 433–460.

Turkle, S. (2010). In good company?: On the threshold of robotic companions. In Y. Wilks (Ed.), *Close engagements with artificial companions: Key social, psychological, ethical and design issues*, (3–10). Amsterdam: John Benjamins.

Turkle, S. (2015). *Reclaiming conversation: The power of talk in a digital age.* New York: Penguin Press.

van Dijk, J. (2018), Designing for embodied being-in-the-world: A critical analysis of the concept of embodiment in the design of hybrids. *Multimodal Technologies and Interaction, 2*(1), 1–21.

Varela, F. J., Thompson, E., & Rosch, E. (1991). *The embodied mind.* Cambridge, MA: MIT Press.

Verbeek, P.-P. (2000). *What things do.* University Park: Pennsylvania State University Press.

Verbeek, P.-P. (2008). Cyborg intentionality: Rethinking the phenomenology of human–technology relations. *Phenomenology and the Cognitive Sciences, 7*(3), 387–395.

Verbeek, P.-P. (2015). Beyond interaction: A short introduction to mediation theory. *Interactions, 22*(3), 26–31.

Vincent, J. (2016, 13 July). Small security robot k5 knocks down toddler, breaks Asimov's first law of robotics. *The Verge.* Retrieved 1 February 2021 from https://www.theverge.com/2016/7/13/12170640/mall-security-robot-k5-knocks-down-toddler.

Volkskrant (2019, 26 September). Een slimme speaker tegen eenzaamheid: Alsof ik er een vriend bij heb. *Volkskrant.* Retrieved 1 February 2021 from https://www.volkskrant.nl/nieuws-achtergrond/een-slimme-speaker-tegen-eenzaamheid-alsof-ik-er-een-vriend-bij-heb~b9810e32/.

Weizenbaum, J. (1966). ELIZA – a computer program for the study of natural language communication between man and machine. *Communications of the ACM, 9*(1), 36–45.

Weizenbaum, J. (1976). *Computer power and human reason: From judgment to calculation.* New York: Freeman.

Wiberg M. (2018) Addressing IoT: Towards material-centered interaction design. In M. Kurosu (Ed.), *Human-computer interaction: Theories, methods, and uman Issues.* HCI 2018. Lecture Notes in Computer Science, vol. 10901. Cham: Springer. https://doi.org/10.1007/978-3-319-91238-7_17.

Zaga, C. (2017). Something in the way it moves and beeps: Exploring minimal nonverbal robot behavior for child-robot interaction. In *Proceedings of the Companion of the 2017 ACM/IEEE International Conference on Human-Robot Interaction* (pp. 387–388). https://doi.org/10.1145/3029798.3034816.

Zaga, C., Lohse, M., Truong, K. P., & Evers, V. (2015). The effect of a robot's social character on children's task engagement: Peer versus tutor. In A. Tapus, E. André, J.-C. Martin, F. Ferland & M. Ammi (Eds), *Social Robotics*, Vol. 9388 (pp. 704–713). Cham: Springer.

6 WHAT CAN ACTOR-NETWORK THEORY REVEAL ABOUT THE SOCIO-TECHNOLOGICAL IMPLICATIONS OF DELIVERY ROBOTS?

Nazli Cila and Carl DiSalvo

The deployment of delivery robots in city infrastructure presents a challenge to companies, municipalities, citizens, designers and academics concerned with the social, ethical and policy implications of urban robotic services. Several companies worldwide are now committed to the increasing use of robots for last-mile delivery, such as food (e.g. Domino's Robotic Unit, Starship Technologies), medicine (e.g. Nuro), goods (e.g. Alibaba and Amazon) and service parts (e.g. thyssenkrupp's TeleRetail). These robots are interconnected, interactive, cyber-physical agents, which can perceive their environment, reason about events and control their actions. Through the use of cameras, sensors and city data, they can navigate the chaos of a city sidewalk and deliver goods efficiently and effectively.

The robotics literature strives to increase the delivery automation by addressing primary technical challenges such as overcoming the physical obstacles at the streets, navigating in traffic or tracking the robot with precision. The deployment of these robots in the real world, however, represented different problems, such as resistance from citizens (e.g. Simon, 2017), vandalism (e.g. Hamilton, 2018) and tort liability (e.g. Yehezkel & Troianos, 2020). Scholars from Science and Technology Studies (STS) and Human-Computer Interaction (HCI) have long pointed out to the urgent gap between systematic representations of social reality embedded in the design of technologies and the messy everyday realities into which they are implemented (Verbeek, 2012; Dourish & Bell, 2013). Correspondingly, designing delivery robots requires a shift from focusing on 'matters of fact', as the robotics field does, to focusing on 'matters of concern', that is, the perceived situations and their consequences (DiSalvo, Lukens, Lodato, Jenkins & Kim, 2014). Treated as a matter of concern, delivery robots include all the technical challenges plus the lived experience and near future effects of delivery automation on individual citizens and society as a whole, together with the value-charged and politicized debates around robotization, job loss, privacy and use of public spaces for profit.

In this chapter, the authors will use Actor-Network Theory (ANT) as an analytical framework to address the matters of concern around the delivery robots and highlight the complex social and ethical issues associated with their deployment in cities. Through ANT's particular lens, a delivery robot can be seen as the materialization of a specific socio-material situation, in which a number of actors each strive to realize their particular idea of what a delivery robot is and could be. ANT is a descriptive, constructivist approach that traces the social and technical relations involved in the development and implementation of new technologies (Callon, 1986; Latour, 1987; Law & Callon, 1992). It follows 'actors', both human and non-human, and the ways they work collectively in 'networks' of action (Latour, 2005, p. 5). ANT does not provide a rigid set of methodological rules for studying networks (Latour, 2005). Rather, it offers a vocabulary for interpretation of the implementation of new technologies (Cresswell, Worth & Sheikh, 2010). This vocabulary will be used in this chapter to ask some critical questions to the delivery-robot-network and thereby identifying some key (current and near future) challenges resulting from the interplay of various actors in it. Although we present a single case, the ANT-based questions are applicable to any other artefact for unravelling the wider ecologies they are embedded in and discussing the sociotechnical implications of their use. In this sense, we offer an approach that proposes a shift from a user-object interaction framework to the notion of increasingly interdependent entanglements between human and non-human actors. This chapter's overall goal is, then, to inspire designers to critically reflect on the consequences and possibilities bound to the growing landscape of intelligence populating our cities and imagine alternative perspectives that can leverage the full potential of the networks they are part of.

In the remainder of this chapter, we will first provide a short introduction to ANT. Then, we will use the primary ANT concepts to present the complex entanglements in the delivery-robot-network. We will conclude the chapter by reflecting on the implications of the conceptual and practical insights we gained from ANT to design.

Actor-Network Theory 101

ANT was developed in the mid-1980s by the sociologists Bruno Latour, Michel Callon and John Law. It can be characterized as a form of relational materialism, as it is concerned with the materials from which social life is produced and the processes by which these are brought into relationship with each other (Prout, 1996). ANT has its own epistemological and ontological position, in essence considering the world as consisting of networks (Law, 1992). The networks can include humans, material objects, concepts – all of which are treated with the same vocabulary. They are referred to as 'actors' (or actants), which have the ability to act and be acted upon. An actor can, however, only act in combination with other actors and in constellations that give the actor the possibility to act (Williams-Jones & Graham, 2003). Thus, inherent to ANT is a move away from the idea that technology impacts on humans as an external force to the view that technology emerged from social interests and that it thus has the potential to shape social interactions (Cresswell *et al.*, 2010). In this sense, ANT presents a substantial reconceptualization of the technology-society relationship. It replaces this dualism with a notion of their mutual constitution and promises a way of understanding devices as participating in and performing social relations alongside human actors (Prout, 1996).

The actor-networks are neither uniform nor stable but ambivalently change over time across social and political contexts, and are the subject of numerous stresses and forces (Williams-Jones & Graham, 2003). They have to be continually maintained through the engagement of the actors involved and may fail and be replaced by other networks. ANT's emphasis on the dynamic and relational aspects of a network is a useful lens for studying non-linear change and the unintended outcomes of technology (Greenhalgh & Stones, 2010). The project plans produced by the management consultants trained in more predictable environments become absurdly inappropriate when, for example, a delivery robot poses a serious safety threat for disabled people crossing the street (Ackerman, 2019) or some citizens set up a candlelight vigil for a delivery robot that caught fire (Lieu, 2018). ANT offers broad and flexible scope for mapping the relevant terrain – the city as a whole with its diverse citizens, experiences, emotions and practices in our case. The central idea of ANT is to investigate and theorize about how networks come into being, trace what associations exist, how they move, how actors are enrolled into a network, how parts of a network form a whole network and how networks achieve temporary stability (Cresswell *et al.*, 2010). To answer these questions, it is essential to consider all the components that collaborate, cooperate, compete and lead to proliferation, persistence or perishing of that network (Williams-Jones & Graham, 2003).

In relation to delivery robots, for instance, the sociotechnical network includes the humans that create, come in contact with and maintain the robot; the city

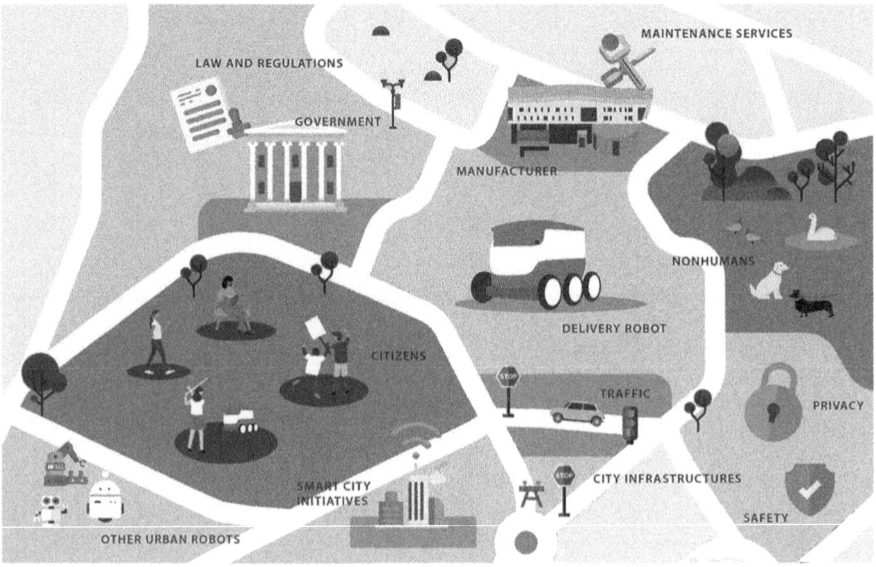

FIGURE 6.1 Actor-network of delivery robots.

infrastructure with its physical affordances, non-human residents and traffic; other technologies which are required for deployment such as sensors, image recognition and tracking; a particular alignment of lawmakers and companies; and abstract concepts such as privacy and citizen rights. Figure 6.1 illustrates this network, but it is by no means exhaustive. The functioning of the delivery robot depends on its being one element within this large sociotechnical network. Each of the human and non-human elements participates in a collective action, which one must mobilize every time s/he calls a delivery robot. In other words, when the delivery robot moves, it is the whole network that moves.

Below, we will draw on the foundational ANT concepts – translation, punctualization, obligatory passage point (OPP), mediator, co-optation and drift – and demonstrate how these concepts can be employed for understanding the power dynamics within an object ecology by using delivery robots as an example.

Delivery robots described through the ANT concepts

Translation

Every actor in a network is essentially independent and capable of resistance or accommodation. There has to be some 'glue' that encourages them to be involved in a network – this glue is *translation* (Williams-Jones & Graham, 2003). Each actor has its own diverse set of interests; thus, a network's stability will result from the continual translation of interests.

Callon (1986) proposed four stages of translation. *Problematization* is about defining the problem and the set of actors who, by defining the problem and the program for dealing with it, make themselves indispensable. *Interessement* has to do with the primary actor(s) recruiting other actors to assume roles in the network. *Enrolment* is about the actors formally taking on these roles. And finally, *mobilization* involves the primary actors to engage others in fulfilling their roles. A hindrance in any of these stages would cease the translation and, therefore, prevent the network to function smoothly.

A useful exercise from the perspective of design would be to employ translation as a means for reflecting on whether the interests of each actor are aligned in a network and whether there are any 'untranslated' parts. This would allow understanding the ecology at play and how an artefact fits within that ecology. Regarding delivery robots, the issue of liability would serve as a good example here. Who is responsible when a delivery robot knocks over some street stands, or worse, a person? Although it is essentially a more complex situation, the primary actors can be simplified as the law, the robot, the manufacturing company and the citizens (problematization). The role of an actor is consolidated and/or redefined during the process of interessement. However, interessement does not necessarily lead to alliances, namely, to actual enrolment. Each entity enlisted by the problematization can submit to being integrated into the initial plan or, inversely, refuse the transaction by defining its identity, goals, motivations or interests in another manner (Callon, 1986). This is where it seems to go wrong in liability cases. Tort law in many legal systems would impose liability on the person or entity in control of a device that causes damage. Unless robots are granted a person-equivalent status (i.e. quasi-persons), somewhat like corporations are now legally recognized as individual entities (see section 'Mediators'), any tortious activity caused by a delivery robot would likely fall on the company that designs, produces or controls the robot (Murphy & Woods, 2009). Yet, with the advances in AI and machine-learning algorithms, robots can take humans completely out of the decision-making loops of the system. This may create situations where the manufacturer/operator cannot be held morally or legally liable for robot's behaviour due to the inability to accurately predict its actions (Barfield, 2018). In this case, determining who is at fault from the robot's actions may be difficult or even impossible under current legal schemes (EU Commission, 2019). Once the legally responsible person for the damage has been identified, to what extent his/her responsibility should be proportional to the 'degree of autonomy' of the robot is the primary question that the current law system is dealing with (Miele & Schiavo, 2018).

To summarize in ANT terms, there has been insufficient translation in solving the liability problem, especially due to the fact that the actors implicated in problematization stage do not acknowledge their roles in this story. Translation allows for identifying the parts of an ecology that fail, and it is in these failures the composition of the ecology become particularly apparent (Law, 1992).

Punctualization

A successful translation will lead to a stable network, at least within the frame of analysis. Each actor's goals and interests become a part of a 'black box' – configurations of actors which have become taken for granted as the way things are and hence are no longer questioned (Akrich, 1992). This is called *punctualization*. At the risk of oversimplification, punctualization can be defined as viewing a combination of actors as one unit.

When networks become stronger and more stable, they can for the purpose of analysis be treated as single points in a larger network (Callon, 1991). In ANT, everything is both an actor and a network; it simply depends on perspective (Cressman, 2009). A delivery robot is a complex network of technology and interactions. This same robot can be also be seen as a single node, punctualized, within a smart city network. One might think of an actor-network as being fractal, expanding infinitely, with each actor being a node in another network (Law, 1999). This perspective is useful for design to understand interactions as being nested in different ways in an ecology and look at the interactions on different scopes and levels.

Although punctualization takes configurations of actors for granted, these black boxes are 'leaky' (Callon & Latour, 1981), meaning that there will always be competing initiatives that seek to open punctualized black boxes (Cressman, 2009). In the network of delivery robots, one of the leaks seems to be about the '1:1 interaction' black box between a human and a robot. These 'single point' framings are common to the HCI and Human-Robot Interaction (HRI) disciplines, which envision an idealized interaction between two essentially isolated actors (Van Oost & Reed, 2010). However, these interactions are quite misleading outside the lab environment, especially when it comes to the urban robotics. When designing robots for public environments, not only is it essential to consider the capabilities of the user, but it is equally important to consider unintended forms of interactions, which involve whoever happens to share the environment with the robot such as passers-by (Salvini, 2018).

But maybe even more importantly, one needs to realize that delivery robots are members of a fleet. They are released to cities in multitudes. These 'swarms' can cooperate as a team and decide on their collective behaviour that emerges from local interactions (Kolling, Walker, Chakraborty, Sycara & Lewis, 2015). The members of swarms can communicate with each other in ways that are not visible/understandable to the human mind. Illah Nourbakhsh (2013) coined the term 'robot smog' to describe an unnerving future where robots will be buzzing all around citizens, taking their photographs and recording their voices, and sharing this information with an interconnected 'massive robot supercolony'. Situations like this could displace the sense of control of the citizens. They can easily end up in situations where they are excluded by the decisions of a group of robots, lacking transparency and accountability.

In ANT terms, the emerging field of robotic swarms poses a challenge for the punctualized interactions between humans and robots. Identifying black boxes and investigating whether there are any threats for their 'depunctualization' would be useful for designers to imagine alternative solutions that could maintain the network or replace them with new black boxes.

Obligatory passage point

An OPP is a critical incident where actors in a network converge around an important issue and the survival of the actor-network is at play (Callon, 1986). The OPPs are often constructed by the primary actor to make itself functionally indispensable to the network (*ibid.*). In other words, an actor will try to structure the network so that the other actors have to pass through it.

Among many other potential critical incidents, an OPP for the delivery robot network could be the necessity for services such as installation, maintenance and repair. Currently, the companies that manufacture the robots provide these services via the human handlers they hire. By this way, the company makes itself indispensable for the smooth functioning of delivery robots in a city. However, it is plausible that other actors may challenge this powerful position. For example, in various cities in the United States, a group of citizens has taken over the task of charging empty electric scooters or fixing the broken ones at their homes during the night. The same situation could well appear regarding the maintenance of delivery robots, in which independent contractors can enter the network as new actors. Or the delivery robots could be designed to seek help from the bystanders when in difficult situations (e.g. Weiss *et al.*, 2015), leverage the capabilities of the 'swarm' it is part of (see section 'Punctualization') and eventually make the maintenance and repair a part of the autonomy of the robot itself.

In ANT terms, new solutions for efficient and cheaper maintenance of delivery robots may come at the detriment of the companies, whose position as an OPP would be challenged. For a network to remain intact, the actors have to adjust their position and dynamics in between (see section 'Translation'). Identifying OPPs is, therefore, a useful exercise for designers to unravel the alliances between multiple actors and their hierarchical power relationships in an object ecology.

Mediators

Latour makes a differentiation between 'intermediaries' and 'mediators'. For translations to develop and network to grow, humans, objects and ideas must act as mediators rather than intermediaries. An intermediary is 'what transports meaning or force without transformation' (Latour, 2005, p. 39). On the other hand, a *mediator* is an actor that makes a difference in the ongoing processes, transforming and translating the meanings in construction (*ibid.*).

A delivery robot can act as mediator or intermediary. When one considers it just as 'a service machine' that delivers food or goods, it comes to act as the latter. However, seeing them exclusively as tools limits the debate around urban robotics to the questions of efficiency and safety, that is, how to design cheaper, better functioning, safer robots? This may impede a more relevant and deeper conversation about the roles and responsibilities of urban robots (Lupetti, Bendor & Giaccardi, 2019). One can also frame a delivery robot as a mediator, who is a contributing member of the urban community, entangled in social relationships, significance and meaning. Such a framing would broaden the scope of the network, introducing interesting concepts such as robot rights and citizenship (Ashrafian, 2015; Rainey, 2016). When seen as a contributing member of a community, citizens may feel responsible to take care of the robots and, in return, expect socially relevant behaviours from the robots. Their relationships transform from being an instrument into start relying on interdependency, collaboration and cooperation.

Mediators act and, as a result, demand new modes of action from other actors (Sayes, 2014). Through this means, they transform, translate and innovate the state of affairs in a network. Exploring how objects can become mediators to transform the ecologies they are embedded in can foreground the questions of meanings and values instead of questions of functionality and thus lead to radical transformations in the technology-driven AI, IoT and robotics fields.

Co-optation

When conditions change, *co-optation* is a key strategy of adaptation for the networks (Dickson, 2000). It allows an actor to obtain certain abilities that it lacks in order to adjust successfully to a new context (Fleron, 1969).

Many cities worldwide have the ambition of becoming a smart city, and co-opting urban robots is one of the means to achieve such a transition. As the cities are the actors-in-power in this discourse, they invite companies to deploy their robotic solutions provided that they make sure these robots 'fit' well into the existing city infrastructures. Therefore, all the research and design efforts go into finding solutions to practical problems such as how to cross streets safely or identify the front doors of buildings. One can, however, also imagine a reverse scenario in which 'the robots co-opt city infrastructures', and the cities are adjusted to fit better to the capabilities of robots. This scenario was demonstrated in a speculative design project involving domestic robots: Rather than attempting to solve complex mechanical problems that commonly become the focal points of research projects, for example, developing a robot hand that can grasp cup handles, the researcher simply redesigned the cup handle in a way that a standard robot hand can hold (Auger, 2014).

The same approach could be adopted in the redesign of cities, where new city architectures are created to better accommodate the robots. In other words, the urban planners and architects would alter the layout of buildings and roads, rather than the engineers alter the robot to cross the street safely. A reframing of the co-optation process can lead to partial resolution of the problems caused by robot-infrastructure match. It would open up a new design space, as well as new and interesting research questions. However, this new frame would also create an OPP, which would change the power dynamics in the network (see section 'Obligatory passage point').

Drift

Drift refers to the transformation of a technology as it is translated into new contexts and used in ways not previously conceptualized by the actors involved in its initial development. The most notable example of drift is where technologies are hacked. The use of robotic drones for smuggling contraband into prison (Albiges, 2019) or smart toys as 'uncontrolled spy devices' (Frenkel, 2017) are recent examples of drift in this regard.

However, drift does not have to be solely about the illegal uses of technology. Technology is ambivalent and can drift as a result of decisions made by many different actors and the need to integrate into pre-existing social and technological contexts (Holmström & Stalder, 2001). When implementing a new technology, it may be necessary to allow it to drift into unexpected situations. If the technology is going to work, it must be open to change (Williams-Jones & Graham, 2003). In the delivery robot network, for example, the consumers may use the robots to lend each other goods or the municipalities can attach routers on the robots to provide public Wi-Fi networks or track the robot data for getting information about sidewalk quality. In all these cases, the initial purpose of the network, which constructed the delivery robots as members of the last-mile delivery of goods, would be subverted by competing purposes such as sharing commodities and improving city services.

The need for drift arises because the emerging actor networks need to be implemented in already existing networks (Holmström & Stalder, 2001). This interaction between old and new creates additional dynamics that are difficult to predict. Hence, adaptive capabilities of all actors, including the technology, are required to deal with unanticipated events. Thinking about drift would help designers to understand the unpredictabilities at play when dealing with multiple agents in an ecology, as well as create opportunities for aligning the multiple interests of multiple actors. If that is successful, the actors in the network will be willing to invest the necessary resources to maintain it because being part of the network serves their individual interests (Williams-Jones & Graham, 2003; see section 'Translation').

Lessons learned from ANT

As aforementioned in the introduction, ANT provides an analytic perspective to produce new descriptions of networks, their qualities and effects. We adopted the ANT vocabulary in this chapter in order to identify the actors in a delivery-robot-network and investigate their dynamic entanglements. This exercise gave us conceptual and practical insights about the implications of ANT to design. First of all, ANT goes beyond the limited perspective of technological determinism which is all too prevalent in the discourse about delivery robots and still common in both the fields of HRI and popular discourses around robotics. The successful functioning of these robots in the city will not happen after solving the technical challenges related to navigation, for example. Nor are we, as either citizens or designers, bound to working within and reproducing contemporary social and political norms. Delivery robots are being embedded in existing and constantly evolving actor-networks of people, ideas, social constructions and evolving norms that are emergent from the existing networks they work within and new networks they cohere together. Thus, they are, to a certain extent, socially shaped and constructed. ANT helps to construct a more nuanced picture of the dynamic relationships between different actors without neglecting their interrelatedness (Cresswell *et al.*, 2010). The value of ANT, then, is that it helps us as designers and design researchers better understand the extent and diversity of relations that bear on delivery robots and on any other new technologies. From this understanding, hopefully, we can better work to shape those relations towards civic ends. Since the robots have been encountering a strong public resistance, ANT's emphasis on studying nonlinear change and the unintended outcomes of technology projects becomes especially important. In mapping the networks of delivery robots, for instance, we might better anticipate what sets of relations might be affected through the introduction of delivery robots. This, in turn, might inform decision making—in when, where, or whether to introduce delivery robots into civic environments.

Furthermore, ANT gave us good questions to ask of the actors in the network and their relations. For example, we looked for the parts of the network that are currently 'untranslated' and why that was the case. We identified the existing the black boxes and assessed if there are any upcoming threats for their depunctualization. We searched for OPPs and deliberated on the power dynamics behind these. We discussed which actors act as intermediaries currently and speculated if there would be any benefits for them to become mediators. We identified which actors have been co-opting to whom and discussed if there are any problems because of this. We found these thinking exercises useful for imagining the diversity of ways that a network might be stabilized, or for that matter, destabilized, or differently configured. For instance, at this moment the world is experiencing a Covid-19 pandemic, in which vast swathes of the global population is subjected to quarantine. Civic actors and actor networks in cities have been differently configured, and albeit in limited capacity, delivery robots are being used for delivering food and medicine in hospitals worldwide. If one were to consider service delivery robots in this moment, in this actor network, one would need to ask again, 'What are the new OPPs? How is co-optation now occurring, or not?' ANT provides us, then, with a pattern for recognizing and describing the multiplicity of networks making up an actor and different visions of how this actor could function or behave.

This brings us to questions concerning the practical value of ANT for design. ANT is considered to be useful for analysing networks that already exist, instead of speculating on the future ones and answering the questions of what if (Law & Hassard, 1999; Kaghan & Bowker, 2001). For this reason, its utility in the context of design is usually considered to be limited (Lindström & Ståhl, 2015; Jenkins, Le Dantec, DiSalvo, Lodato & Asad, 2016). This makes sense as ANT is borne of the social sciences, rather than design, and so on its own ANT is not generative in the ways that many design methods are generative. Still, in our experience as design researchers we were able to use ANT to assist in imagining alternative situations, especially after identifying untranslated parts of the network, OPPs and black boxes. That is, the practical value of ANT to design research is that ANT provides a technique for enumerating the actors, their qualities and relations within a field of inquiry. When those actors, qualities and relations are listed, they can be used as the material for all manner of generative methods within design as the basis for inventive and critical making. So, it is not that ANT is generative, but that it helps designers produce the materials to be generative with. What is particularly important is those materials are relational and express diverse agencies. So it is not the will of the designer alone but, rather, the designer taking into account the animated character of more-than-humans actors within a network. Correspondingly, once we identified these actors and their entanglements, ANT allowed room for imagination on potential solutions and alternative futures, borrowing from the emerging topics in HCI and HRI such as robot citizenship, swarms, more-than-human cities, robot responsibilities and citizenship.

One might ask how an ANT-informed research approach is different from stakeholder mapping, which is one of the staple methods of designers in the pre-ideation stage of technology development. ANT assumes a radical symmetry between human and non-human actors. This is profoundly different from many approaches to design, most notably, human-centred or user-centred design. Contrary to all human actors that are typically addressed in a stakeholder map, in an ANT network humans made comparable in status to non-humans, such as a sidewalk, a pigeon or legislation. And just as importantly, the sidewalk, pigeon or legislation is made comparable to the human in terms of their effects on configuring agency and what we call 'society'. This flat ontology was criticized on various occasions (Mutch, 2002), but ANT does not imply or require that all entities be treated as identical for all purposes, not that the various relations between actors be egalitarian (Williams-Jones & Graham, 2003). In fact, the benefit of laying out a network is tracing the types of relations, interactions and dependencies among actors, and determining the flow of power and controls. For this reason, ANT provides a richer overview of the design context and offers practical insights to designers. Christiano Storni (2015, p. 169) asserts that ANT helps designers to 'look at the design task not as what entities are in isolation, but rather what they become, do and produce when they are associated together'.

Conclusions

ANT can destabilize the dominant discourse around a matter of concern. By unpacking that which has been simplified, a rich, complex understanding of a case develops that enables sustained social critique (Williams-Jones & Graham, 2003). As has been shown in our deliberations on sharing urban spaces with delivery robots, the wider diffusion of these robots in the cities raises social and ethical concerns about the liability of the robots for the damage they may cause, the potential intimidation caused by the 'robot smog', providing service and maintenance, the role and responsibilities of robots in the communities, designing specialized infrastructures for the robots to function effectively and dealing with unintended uses of delivery robots. By applying the concepts of translation, punctualization, OPP, mediator, cooptation and drift to the development and implementation of delivery robots, it was possible to better see the complex and multifaceted nature of the networks in which they are situated. The resulting overview can help account for interrelational interests of various actors in this network and enable a more nuanced and comprehensive analysis in this regard. In this, our hope is that this chapter can provide the interaction design community with an approach to critically reflect on and rethink the design of robots, and any other smart object in that matter, in meaningful and responsible ways.

Bibliography

Ackerman, E. (2019, 19 November). My fight with a sidewalk robot *[Blog post]*. Retrieved 26 February 2021 from https://www.citylab.com/perspective/2019/11/autonomous-technology-ai-robot-delivery-disability-rights/602209/.

Akrich, M. (1992). The de-scription of technical objects. In W. E. Bijker & J. Law (Eds), *Shaping technology/building society: Studies in sociotechnical change* (pp. 205–224). Cambridge, MA: MIT Press.

Albiges, M. (2019, October 14). Dozens of drones spotted hovering near Virginia prisons. *Government Technology*. Retrieved 26 February 2021 from https://www.govtech.com/public-safety/Dozens-of-Drones-Spotted-Hovering-Near-Virginia-Prisons.html.

Ashrafian, H. (2015). Artificial intelligence and robot responsibilities: Innovating beyond rights. *Science and engineering ethics, 21*(2), 317–326.

Auger, J. (2014). Living with robots: A speculative design approach. *Journal of Human-Robot Interaction, 3*(1), 20–42.

Barfield, W. (2018). Liability for autonomous and artificially intelligent robots. *Paladyn, Journal of Behavioral Robotics, 9*(1), 193–203.

Callon, M. (1986). Some elements of a sociology of translation: domestication of the scallops and the fishermen of St Brieuc Bay. *The Sociological Review, 32*(1_suppl), 196–233.

Callon, M. (1991). Techno-economic networks and irreversibility. In J. Law (Ed.), *A sociology of monsters: Essays on power, technology and domination* (pp. 132–161). London: Routledge.

Callon, M., & Latour, B. (1981). Unscrewing the big Leviathan: How actors macro-structure reality and how sociologists help them do so. In K. Knorr-Cetina & A.V. Cicourel (Eds), *Advances in social theory and methodology: Toward an integration of micro- and macro-sociologies* (pp. 207–303). Boston: Routledge & Kegan Paul.

Cressman, D. (2009). A brief overview of actor network theory: Punctualization, heterogeneous engineering & translation. Vancouver: ACT Lab/Center for PolicyResearch on Sicience & Technology (CPROST) School of Communication, Simon Fraser University (Working Paper).

Cresswell, K. M., Worth, A., & Sheikh, A. (2010). Actor-network theory and its role in understanding the implementation of information technology developments in healthcare. *BMC Medical Informatics and Decision Making, 10*(1), 67.

Dickson, B. J. (2000). Cooptation and corporatism in China: Thelogic of party adaptation. *Political Science Quarterly*, 115, 517–540.

DiSalvo, C., Lukens, J., Lodato, T., Jenkins, T., & Kim, T. (2014, April). Making public things: How HCI design can express matters of concern. In *Proceedings of the SIGCHI Conference on Human Factors in Computing Systems* (pp. 2397–2406). https://doi.org/10.1145/2556288.2557359.

Dourish, P., & Bell, G. (2013). Resistance is futile: Reading science fiction alongside ubiquitous computing. *Personal and Ubiquitous Computing, 18*, 769–778.

EU Commission (2019). Liability for artificial intelligence and other emerging technologies. Retrieved from https://ec.europa.eu/transparency/regexpert/index.cfm?do=groupDetail.groupMeetingDoc&docid=36608.

Fleron, F. J. (1969). Cooptation as a mechanism of adaption to change: The Soviet political leadership system. *Polity, 2*, 176–201.

Frenkel, S. (2017, December 21). A cute toy just brought a hacker into your home. *New York Times*. Retrieved 26 February 2021 from https://www.nytimes.com/2017/12/21/technology/connected-toys-hacking.html.

Greenhalgh, T., & Stones, R. (2010). Theorising big IT programmes in healthcare: Strong structuration theory meets actor-network theory. *Social Science & Medicine, 70*(9), 1285–1294.

Hamilton, I. A. (2018, June 9). People kicking these food delivery robots is an early insight into how cruel humans could be to robots. *Business Insider*. Retrieved 26 February 2021 from https://www.businessinsider.com/

people-are-kicking-starship-technologies-food-delivery-robots-2018-6?international=true &r=US&IR=T.

Holmström, J., & Stalder, F. (2001). Drifting technologies and multi-purpose networks: The case of the Swedish cashcard. *Information and organization, 11*(3), 187–206.

Jenkins, T., Le Dantec, C. A., DiSalvo, C., Lodato, T., & Asad, M. (2016, May). Object-oriented publics. In *Proceedings of the 2016 CHI Conference on Human Factors in Computing Systems* (pp. 827–839). New York: ACM.

Kaghan, W. N., & Bowker, G. C. (2001). Out of machine age?: Complexity, sociotechnical systems and actor network theory. *Journal of Engineering and Technology Management, 18*(3–4), 253–269.

Kolling, A., Walker, P., Chakraborty, N., Sycara, K., & Lewis, M. (2015). Human interaction with robot swarms: A survey. *IEEE Transactions on Human-Machine Systems, 46*(1), 9–26.

Latour, B. (1987). *Science in action: How to follow scientists and engineers through society*. Milton Keynes: Open University Press.

Latour, B. (2005). *Reassembling the social: An introduction to actor-network-theory*. New York: Oxford University Press.

Law, J. (1992). Notes on the theory of the actor-network: Ordering, strategy, and heterogeneity. *Systems Practice, 5*(4), 379–393.

Law, J. (1999). After ANT: Complexity, naming and topology. In Law, J. & Hassard, J. (Eds), *Actor network theory and after* (pp. 1–14). Oxford: Blackwell.

Law, J., & Callon, M. (1992). The life and death of an aircraft: A network analysis of technical change. In W. E. Bijker, & J. Law (Eds), *Shaping technology/building society: Studies in sociotechnical change* (pp. 21–52). Cambridge, MA: MIT Press.

Law, J., & Hassard, J. (Eds). (1999). *Actor network theory and after*. Oxford: Blackwell.

Lieu, J. (2018, 17 December). Delivery robot catches fire at university campus, students set up vigil. *Mashable*. Retrieved 26 February 2021 from https://mashable.com/article/kiwibot-fire-uc-berkeley/?europe=true.

Lindström, K., & Ståhl, Å. (2015). Figurations of spatiality and temporality in participatory design and after–networks, meshworks and patchworking. *CoDesign, 11*(3–4), 222–235.

Lupetti, M. L., Bendor, R., & Giaccardi, E. Robot citizenship: A design perspective. In S. Colombo, M. Bruns Alonso, Y. Lim, L-L. Chen & T. Djajadiningrat (Eds), *Design and Semantics of Form and Movement* (pp. 87–95). Cambridge, MA: MIT Press.

Miele, C. O., & Schiavo, V. (2018, December). Robots and liability: Who is to blame? *Dentons*. Retrieved 26 February 2021 from https://www.dentons.com/en/insights/articles/2018/december/20/robots-and-liability.

Murphy, R., & Woods, D. D. (2009). Beyond Asimov: The three laws of responsible robotics. *IEEE Intelligent Systems, 24*(4), 14–20.

Mutch, A. (2002). Actors and networks or agents and structures: Towards a realist view of information systems. *Organization, 9*(3), 477–496.

Nourbakhsh, I. R. (2013). *Robot futures*. Cambridge, MA: MIT Press.

Prout, A. (1996). Actor-network theory, technology and medical sociology: An illustrative analysis of the metered dose inhaler. *Sociology of Health & Illness, 18*(2), 198–219.

Rainey, S. (2016). Friends, robots, citizens?. *ACM SIGCAS Computers and Society, 45*(3), 225–233.

Salvini, P. (2018). Urban robotics: Towards responsible innovations for our cities. *Robotics and Autonomous Systems, 100*, 278–286.

Sayes, E. (2014). Actor–network theory and methodology: Just what does it mean to say that nonhumans have agency?. *Social Studies of Science, 44*(1), 134–149.
Simon, M. (2017, June 12). San Francisco just put the brakes on delivery robots. *Wired.* Retrieved 26 February 2021 from https://www.wired.com/story/san-francisco-just-put-the-brakes-on-delivery-robots/.
Storni, C. (2015). Notes on ANT for designers: Ontological, methodological and epistemological turn in collaborative design. *CoDesign, 11*(3–4), 166–178.
Van Oost, E., & Reed, D. (2010, June). Towards a sociological understanding of robots as companions. In *International Conference on Human-Robot Personal Relationship* (pp. 11–18). Berlin: Springer.
Verbeek, P-P. (2012). Expanding mediation theory. *Foundations of Science, 17*(4), 391–395.
Weiss, A., Mirnig, N., Bruckenberger, U., Strasser, E., Tscheligi, M., Kühnlenz, B., *et al.* (2015). The interactive urban robot: user-centered development and final field trial of a direction requesting robot. *Paladyn, Journal of Behavioral Robotics, 6*(1), 42–56.
Williams-Jones, B., & Graham, J. E. (2003). Actor-network theory: A tool to support ethical analysis of commercial genetic testing. *New Genetics and Society, 22*(3), 271–296.
Yehezkel, A. & A. W. Troianos (2020, February 23). Legal considerations before deploying autonomous delivery robots. *The Spoon.* Retrieved 26 February 2021 from https://thespoon.tech/legal-considerations-before-delploying-autonomous-delivery-robots/.

PART THREE

METHODOLOGIES

7 SKETCHING AND PROTOTYPING SMART OBJECTS

Philip van Allen

Artificial intelligence (AI) and machine learning (ML) are unique design materials and are a challenging fit for conventional design methods. The black-box character of 'smartness' makes designing a smart object especially enigmatic. This makes the thoughtful choice of design methods especially important.

This chapter explores how students, designers and researchers can approach smartness as a design material (Redström, 2005), focusing on explorations, sketching and prototyping smartness, hands-on methods, tools and practitioners' notes.

Sketching and prototyping

Sketching and prototyping are key strategies for the design of smart objects. Sketching is fast, provisional, low cost, productively ambiguous and exploratory, whereas prototyping is often a longer, more detailed process focused on refining towards a final design (Buxton, n.d.). Both can be challenging when designing smart objects because of the unfamiliarity with, and time involved in, creating functional AI systems (Table 7.1, *ibid.*).

Because each designer has their own definitions, it is important to be clear about the different approaches to designing and how they address the challenges outlined in this chapter. In 'Sketching User Experiences', Buxton (2010) argues that there is a continuum of intents that begins with sketching (getting the design right) and finishes with prototyping (getting the right design), while important strategies for conventional domains, sketching and prototyping for smart objects have extra challenges.

Table 7.1 Bill Buxton, sketching to prototyping continuum. What sketches (and prototypes) are, and are not.

Continuum of Design Intents	
Sketch ——————>>>	**Prototype**
Invite	Attend
Suggest	Describe
Explore	Refine
Question	Answer
Propose	Test
Provoke	Resolve
Tentative, non-committal	Specific Depiction

The study of human-AI interaction design

The challenges for designing AI systems are an increasingly studied area in Human-Computer Interaction (HCI) (Yang, Steinfeld, Rosé & Zimmerman, 2020) Researchers have identified several challenges for designers:

- Unfamiliarity with the capabilities of AI
- Unpredictability of AI outputs
- Challenges in iterative prototyping and testing human-AI interaction
- Traditional human-centred design processes don't always work well for autonomous, smart objects that interact with humans, data, the environment and other smart objects

Getting started

If smartness is a new design material, designers need to be able to effectively sketch and prototype to create the most interesting and appropriate outcomes for a particular project. How to do this depends greatly on the questions of WHO, WHY, WHAT and HOW.

To successfully begin the design process for a smart object, it is critical to align your approach and methods with your type of object being designed, the team's skill sets and project goals. At the beginning, the design team must define the following for the project, so the sketching and prototyping processes can take these project characteristics into account.

WHO: Who are the team members and what are their skill sets and interests? Domain experts, creative technologists, data scientists, business people, researchers or interaction designers? Expertise in these different disciplines should lead to different approaches that fit the members of design team. For example, a ML-based medical diagnosis project may rely heavily on a doctor's expertise to work with a data scientist to define ML training data quality. And a creative technologist might put together a quick sketch of how the interaction might work using a small sample data set of X-rays provided by the business defined source.

WHY: What is the purpose of the sketch or prototype? Is it to identify opportunities, learn the medium, brainstorm, explore a design space, ideate, test usability, refine a design for production and release or more broadly to consider ethics and potential undesirable outcomes? Defining the 'why' will lead to a more helpful brief for the sketch or prototype.

WHAT: What kind of object is being designed? Is it an interface on a screen? Is it a physical device? How will the smart object and human interact? Through voice, touch, gesture, text, or brain-machine interface?

HOW: What strategies and methods will the designers use? WoZ, Marionetting, Visual Tools, hand-coded Python? These can significantly affect the overall outcome, cost and timeline. To help determine the how, the team may initially choose to do some tinkering with the technology, or interview data scientists with expertise in the project domain.

Smartness as a design material

As designers, we should consider the character and affordances of smartness just as we might working with any other design material. Different smartness algorithm and the diverse human intelligences we might model digital smartness on (Visual-Spatial, Linguistic-Verbal, Logical-Mathematical, Bodily-Kinesthetic, Musical, Interpersonal, Intrapersonal, Naturalistic) have very different qualities (Gardner, 1983). Specifically, as we shape the content, intent and behaviour of our project, we should think about the grain of the material (i.e. its inherent qualities). Is the design working with or against the material properties? What design opportunities does the material present? What could go wrong? How smart is it in the intended domain(s)? How can any particular kind of digital smartness be creatively abused? Can it be trusted? What should its perceived personality be in this context? There are several challenging and productive characteristics of smartness: unpredictability, errors and emergence; unintended outcomes; contextual adaptation; multimodal communications; animism; and social relations.

Unpredictability, errors and emergence

Smart objects depend on complex algorithms (neural nets, behaviour trees, fuzzy logic, intentional randomness, reinforcement learning, etc.) that produce unpredictable and often unexplainable behaviours. As such, the outputs of AI systems are 'non-deterministic' and the very strengths of useful intelligence (adaptability, creativity, learning, inference, problem-solving, serendipitous opportunism, contextual judgement, goal balancing, risk taking and errors) tend to violate the common tenets of UX design such as consistency, transparency, avoidance of errors and efficiency.

This complexity and unpredictability also make it hard to assess how the object will live in the world, especially while interacting with other autonomous smart things such as humans and objects.

Unintended outcomes

The emergent qualities of tangled smart object ecologies create a special responsibility for designers to speculate about the range of possible outcomes and preventing the worst cases. As an example of the caution designers need in the face of over-optimism for AI, there is growing scepticism about the future success and safety of autonomous vehicles (AVs), which are promoted to improve traffic and safety. For example, urban planner Adam Millard-Ball (2018) applied game theory to AVs and came to the following observation: 'Because autonomous vehicles are by design risk-averse ... pedestrians will be able to act with impunity ... adoption of autonomous vehicles may be hampered by their strategic disadvantage that slows them down in urban traffic' (Lasnier, 2016). In other words, despite the good intentions for AVs, pedestrians (not to mention human drivers) may 'game' the AVs, knowing the vehicles are risk-averse. This may result in the unintended outcome that traffic will become worse because of AVs.

Contextual adaptation

To continue with the driverless car example, because each city has different driving cultures, AVs may have a difficult time behaving safely. Imagine the driving differences in Los Angeles, London, Mumbai and Berlin. To take advantage of smartness, objects must be designed to adapt to their context, and this creates additional design complexity and opportunities for failure.

Multimodal communications

Smart objects are challenging to design in part because they often use a multimodal communication capability with humans and other systems. This involves designing interactions that drift across screens, voice, 'telepathy' (i.e. brain-computer interfaces (BCIs) or wireless vehicle-to-vehicle communication (V2V)) and tangible interaction. The designer must accommodate fluid mode shifts, while clearly indicating the current mode to a person. For example, is this brand of AV telling a pedestrian or cyclist to go ahead by blinking its lights, by speaking with a voice? Or both? Does the communication style vary from city to city?

Animism

When things behave on their own, humans tend to ascribe life and intention to them (Urquiza-Haas & Kotrschal, 2015), whether the designer intended it or not. While this may seem to be an irrational or naive human trait, designers should consider that life and 'personality' can be used as a design affordance to help make the smart object more understandable and interesting (van Allen & Marenko, 2015).

This use of animistic design leverages human habits that use animism as a metaphor to predict behaviour through a theory of mind for the smart object – the human may predict that 'alive' things typically behave a particular way because of empathetically guessed intentions. Designing this perceived animism can be a powerful way for designers to make the inherent biases and limits of smart things explicitly and more honestly represented.

Animism also allows for the creation of diverse smart things, each of which has a different point of view. Because of this, animistic design has the advantage of indicating to humans that smart things are diverse and fallible. It may be better to instil an intentional level of distrust in the smart things we design because this is safer for humans. But there is a skeuomorphic trap here as well – it is easy to use anthropomorphic tropes like cuteness and emotive faces as a seductive crutch which may communicate more intelligence, values and empathy than are actually in the smart object. The challenge is to use subtle design cues and aesthetic choices to clearly indicate the limits and capabilities of the 'dumb smart' object. There is also a danger in creating a captain-conscript relationship between the human and smart object, which may lead to habits of negative social behaviour.

Social relations

Productive smartness needs conversation and interaction to gather information and context. And with the animistic impression people get of smart things, they tend to include them in their social context, as they do with a pet. Designing with the social in mind means embracing the subtle aspects of social interaction – inflection, empathy, interpersonal history, nonverbal cues, greetings, goodbyes, trust and so forth.

Strategies

There are many approaches that can work for designing AI systems. As discussed, the method chosen depends on the goals for the current stage of the project as well as the skills of the team. While there are challenges, some conventional design methods will continue to work for AI-based projects, while realizing the limitations of these methods for AI systems (van Allen, 2017). In particular, the autonomous nature of smart objects means that they are not only interacting with a 'user' but they also become part of an ecology with a mixed population of humans and other smart objects. For this reason, 'human-centred' design principles may break down because creation of successful smart systems often requires the decentring the human while understanding that smart objects and the ecologies they participate in have their own 'needs' and perspectives.

This is not to say that smart objects should gain any particular privilege over humans (Birhane & van Dijk, 2020). Instead, designers should design for the success of the milieu for the benefit of all the participants. 'Happy' participants make for a better party. Healthy participants make for a better ecology.

At the early stages of the design process, it is important for the team to engage in a divergent process of design ideation and discovery. This includes not only a process of UX research but also exploratory sketching and some 'tinkering' with data and technology that enable the team to gain a deeper understanding of smartness as a design material. As the design concepts mature, it will become necessary to create a working prototype of the system to allow for critique and usability testing.

Exploring

New design materials and technologies often require a more exploratory, critical and provocative approach for developing effective project concepts – to help avoid clichés and hype-driven 'solutions'. In this section, we'll discuss approaches that can help the designer break out of their assumptions and preconceptions about smart objects.

Critical prototyping

In my design practice and teaching, I've tried several approaches, which typically involve a defamiliarizing technique. The following are briefs for what I call 'critical prototyping', a concept development approach that applies to a thinking-through-making strategy.

Useless AI: In some industrial design curricula, there is a tradition of creating useless products, for example, a toothbrush that doesn't work (https://www.

theuncomfortable.com; Shopikon, 2014). This approach helps question the purpose of the object being designed, while challenging the conventions of functional design. This upside-down approach forces the designer to reconsider their assumptions. And taking a more speculative and critical direction requires an openness on the part of the designer, client, student and technologist to the unconventional.

> Brief: Design a smart object that takes a critical perspective of AI/ML by designing something that on the surface seems plausible and sensible but on deeper analysis is useless, absurd or off base in some revealing way.
>
> The term 'useless' is open to broad interpretation, but your project must take a position and dig into the challenges, affordances, unforeseen side-effects and potential failures of AI and ML. Your project should be grounded in insights drawn from real ML experiments. To succeed, you must take risks with a sense of criticality and humour.

In a studio focused on AI called the 'Internet of Enlightened Things, AI in the Neighborhood' (Hooker & van Allen, 2017), the Useless AI prompt led to several excellent starting points for students (who had no prior experience in AI). After the studio was completed, the class participated in Ars Electronica in the fall of 2017 and contributed to the growing discourse around IoT (Frank, Pührerfellner, von Rechbach & Lechner, 2017).

Strange new creatures: This approach is intended to get designers to rethink smart systems and how they will live in the world. AI systems are inherently strange because of their mix of human parentage, algorithmic quirks and a kind of savant syndrome (Treffert, 2014) with particular 'islands of genius' areas of expertise. With a strange backstory, the designer is forced to think about how a smart system behaves, interacts and (seems to be) motivated.

> Brief: Come up with a strange new creature to model your project on. This creature should have an exotic backstory. For example, it might be from another planet, or it lives miles under the sea, or its job is to work on arcane math problems. Based on this backstory, come up with the following:
>
> - How does it express itself?
> - What are its goals/mission?
> - What are its senses and how does it perceive?
> - What is its form and how does it behave?
> - What are its data sets and biases?
> - What qualities/values/biases did it inherit from its makers?
> - What is its personality?
> - What are its skills and limits?

Then put a human in the speculative mix, spin the critical blades and imagine what happens when people interact with the strange new creature:

- What are the misunderstandings between the human and the robot?
- How do the human and the robot bridge their gaps? In communication, values/ethics, goals, perception, mental models?
- What sorts of ecologies and outcomes emerge with a community of strange creatures mixed with humans?

Science fiction prototyping: This approach uses a work of science fiction as a prompt for designing a project. Developed as a course by Sophia Brueckner, it fits into Sophia's notion of critical optimism (STAMPS, n.d.), which strikes a productive balance between techno-optimism and dystopic critical design.

> Brief: For decades, science fiction authors have explored both our wildest dreams and greatest fears for where technology might lead us. Yet, science fiction is fuelled by the concerns of today just as much as it is about fantastic imaginings of the future. This class ties science fiction with speculative/critical design as a means to encourage the ethical and thoughtful design of new technologies.
>
> With a focus on the creation of functional prototypes, this class combines the analysis of classic and modern science fiction texts and films with physical fabrication or code-based interpretations of the technologies they depict.

Tinkering: It takes focused technology tinkering for designers to achieve a sketching and prototyping level of understanding of AI as a material of design, especially in terms of its affordances, dangers and limits. If the person doing the sketching or prototyping does not have an AI or data science background, it is best to start with some informal experiments with AI/ML technologies. This kind of tinkering in technology can provide the at-hand, tacit understanding (Schön, 1984) necessary for great design.

In my experience as an educator, designer and technologist, tinkering occupies a productive place outside of Buxton's sketching/prototyping dichotomy. I also find that implementing a rough, working sketch in an actual technology (not a 'prototype') helps in educating the designer and stakeholders about the grain of the material – how to productively work with and against it. This is especially true when working with or inventing design for unfamiliar technologies such as extended reality (XR) and AI. For example, give yourself a small tinkering project:

- Hack together a basic recognition example from free tools such as Google's Teachable Machine; for example, try using any of the three different input types – hand signals, sounds or body poses.

- Modify a sample chatbot from a free trial version of IBM's Watson Assistant.
- Use one of your own data sets (say, your pet pictures) to train an ML system such as Google's Teachable Machine.

In addition these technical self-assignments, it is important to experiment with contemporary AI/ML services such Amazon Alexa, Google Assistant and Apple Siri. Similarly, long-term playing with commercial smart objects such as the iRobot Roomba vacuum and the Google Nest thermostat can shed light on important qualities of smart objects that live in the real world.

Readings

To gain a deeper understanding of design approaches to AI, the designer should review recent works published in venues such as SIGCHI, AAAI and by the community of commercial and non-profit organizations working to develop thoughtful approaches to the design of AI.

A selected bibliography follows:

An AI Pattern Language, Data & Society – https://datasociety.net/output/ai-pattern-language/

The Copenhagen Letter, Copenhagen TechFestival – https://copenhagenletter.org

Guidelines for Human-AI Interaction, Microsoft – https://www.microsoft.com/en-us/research/publication/guidelines-for-human-ai-interaction/

AI Now Institute – https://ainowinstitute.org

Montreal AI Ethics Institute – https://montrealethics.ai

The Asilomar AI Principles – https://futureoflife.org/principles-discussion/

Google Clips, AI product case study – https://design.google/library/ux-ai/

Paper summarizing global AI Ethics positions – https://arxiv.org/pdf/1906.11668.pdf (Gonfalonieri, 2018)

Jason Mayes – Machine Learning 101

Gene Kogan – Machine Learning for Creatives Video

Sketching in AI

Sketching is often skipped in technological contexts, where it can seem strange or difficult to use hardware and software as one might use a sketchpad and pencil. Even so, it is important to do the kind of informal and propositional ideation one associates with conventional sketching. In recent years, many entrepreneurs, designers and engineers have adopted the strategy of sketching in technology. They do so in the sprite of the maker movement, using a wide range of open-source software/hardware tools. This movement has been supported for the last fourteen years by the annual Sketching in Hardware (Kuniavsky, 2014) conference organized by Mike Kuniavsky and Tod Kurt. In one description of the event (Dore, 2009):

> Hardware sketches are the tools or building blocks of technology design. They allow the designer to explore experiences mediated by products or staged in spaces without requiring engineering support during creative phases. If a sketch of a static device can be thought of as a noun, a sketch of an electronic device must be closer to a verb. So while a designer can create storyboards to determine whether a phone should vibrate under specific conditions, like the intensity of light in a given space, to get a feeling for what that really means, a working device—a sketch model—needs to be built.

A quote about the 2017 conference concisely expresses the sketching in hardware attitude (Carpenter, 2017): 'Prototypes are questions embodied.'

Smart objects are typically composed of a mix of hardware and software, which affect the qualities of the device behaviour. As this book posits, we will soon live our everyday lives with smart objects. To design these everyday interactions, designers need to experience and test in an embodied and tangible way, which is what an AI sketch affords. For example, in real life, how does it feel to be talking to a driverless car in a traffic jam? An AI sketch could explore many of the factors that should be considered in designing such as system.

Sketching in AI involves creating quick, exploratory designs that can be provocative, experimental and surface new questions. These quick design sketches are intentionally rough so designers and other stakeholders do not become overly committed to any one design direction. The challenge of this design strategy is that without a full ML implementation, it is difficult to simulate the unpredictable and unusual qualities of ML algorithms. Sketching is also a social activity, in that it creates new conversations while creating a community around the ideas embodied in the sketch. There are several approaches to sketching in AI:

- **WoZ**: Wizard of Oz technique. In this approach, the team fakes the behaviour of a system as a substitute for making a working prototype.
- **Marionetting**: Related to WoZ, this strategy replaces the smartness algorithms with a human who manipulates the behaviour or the system as if it were smart (Wang, Sibi, Mok & Ju, 2017). The system is designed with hidden 'strings' or controls so that a person can work the system

like a puppeteer, reacting to a user's interaction in real time. For example, triggering certain preprogrammed 'behaviour macros' or literally 'voicing' the speech output of the envisioned system.

- **Visual tools**: There are a growing set of tools being developed to make the sketching of AI systems easier by eliminating the need to write code to create a working system. These tools allow the designer to create a partially working system that can be used for sketching, reflection and experimentation.

Prototyping AI

As a smart object concept comes together after doing the appropriate research, exploration and sketching, the design will need to be refined and tested so the team can further iterate the design. This leads to a process of prototyping working versions of the concept. This prototyping can be done with a range of tools and may need the help of a data scientist and/or programmer. It may be useful to use the following process:

1. Acquire and process the dataset for training the ML model.
2. Test the resulting trained model to see how it behaves with 'real-world' inputs.
3. Work with a programmer/creative technologist to build a quick working demo and do some quick testing.
4. Work with a data scientist to refine and optimize the model based on the testing results.
5. Update the UX to address concerns found in testing and critique.

As mentioned, developing a fully functional AI prototype is challenging and requires significant expertise in data sets, ML training and ML model optimization. It may also involve selecting a platform to run the project on.

Edge computing

ML models can sometimes be run 'at the edge', meaning that the device (be it a phone, camera or other system) has the computational power to run AI at a reasonable speed. This is even possible on low-cost maker tools such as the Raspberry Pi. The benefit of AI at the edge is that any data collected for the smart thing stays on the device and does not have to be shared with an outside party. So for contexts where privacy is critical, it may be important to be able to state that the photos or conversations never leave the device.

Cloud computing

In contrast to edge computing, cloud computing makes much more powerful computational capacity available to a smart object. In this case, any data being processed by the smart object is uploaded and analysed in the cloud on remote servers. These severs could be owned by the smart object company, or they could be on rented servers such as Amazon Web Services (AWS) where the software is proprietary to the specific smart object. Alternatively, the ML models could be a software as a service (SaaS) such as IBM Watson, where the company pays the service to use their models to perform the necessary ML tasks. Here the data (such as a voice recording or camera image) is uploaded from the device to the SaaS where it is processed by their servers and their ML models. Different ML SaaS vendors have different privacy policies that should be examined to determine if they are appropriate for the application.

Data sets and ML training for a prototype

Creating a working AI system sufficient for a full prototype requires collecting a quality data set with labelled elements and building a fast and accurate ML model trained on this dataset.

1. **Obtain the dataset**: Be careful to enforce privacy, data security and copyright policies. Implement data collection strategies to ensure minimal bias. Document the entire process to give the training data sufficient provenance to be later analysed.
2. **Clean the data**: Remove or fix bad data.
3. **Label the data**: Assign labels to the data so the ML system can differentiate between different inputs, for example, cat/dog, cancer/benign and so forth
4. **Train/test the model – 80/20 rule**: Use about 80 per cent of the data to train the ML system and create a model, and then use the remaining 20 per cent of the data to test the model for accuracy with 'fresh' data (don't use training data for testing, as that would not be objective since the model has already 'seen' this data).
5. **Optimize the model**: Make adjustments to the model and training process to make the model more accurate and faster.
6. **Test the model and repeat** the process to improve performance.

ML process considerations

Collecting, cleaning and labelling data: The old saying, 'garbage in, garbage out', is particularly apt for ML, especially when considering issues such as unintended bias. If the data collection and labelling methods are not well designed, biases may creep into the data and any ML models trained on that data.

In addition, the data are often less than ideal or accurate. For example, in IoT contexts, sensors measuring characteristics such as temperature may be 'noisy' and have significant errors due to the nature of the sensor itself or the design of the system. These problems have to be cleaned up before the data is ingested into the ML model through training. This clean-up is sometimes a tedious manual process and other times performed algorithmically (e.g. applying a smoothing function to time series data).

ML algorithms: Select an appropriate ML system technique that works well with the data and project goals. Certain kinds of data and ML tasks are better accommodated with specific ML algorithms.

Training, testing and optimizing: Achieving the desired ML model requires a careful process of training the model with the data set as input, testing the resulting model against a 'fresh' data set that is separate from the original training data and 'tuning' the model so it performs at an optimum level from an accuracy and speed perspective.

To explain the process, let us imagine you are going to create a system that identifies people's gender through facial recognition.

Gather the training data set

Developing a large data set for training an ML system (e.g. ten thousand photos of faces labelled with names and genders) is time-consuming and full of risks. Assembling the data set requires careful collection and labelling practices, therefore someone with expertise in data science should supervise the compilation of the dataset.

Clean the data

The resulting data set must be 'clean' where the elements (whether facial images or temperature sensor data) must be of high quality; include minimal 'bad' data (e.g. blurry images or inaccurate sensor values), mistaken labelling or inappropriate biasing (note that some bias may be inherent in the context).

For example, with a face image identification system described above, a process of 'cleaning' is required to eliminate non-face images, screening the images to ensure a diverse range of skin tones, gender identities, hair styles and accessories (e.g. glasses). Further, the images must be labelled appropriately and 'accurately' in a way that will work best with the kind of 'real-world' images and requests that will be put through the face identification system.

Alternatively, one can use an existing labelled data set such as ImageNet (http://www.image-net.org), an open-source collection of 14,197,122 labelled images maintained by Stanford University, or WordNet (https://wordnet.princeton.edu), a large lexical database of 155,327 English nouns, verbs, adjectives and adverbs.

These have their own challenges, since training a successful ML model with such a large data set can consume significant computational power, time and a high degree of technical expertise and experience. It is important to recognize that while ImageNet has been widely used in the AI community, it is deeply flawed (Crawford & Paglen, 2019; Prabhu & Birhane, 2020). Because assembling and processing a data set can be so labour-intensive and expensive (which can inhibit the design process), it is useful to apply the notion of minimum viable product (MVP, from the Agile design process) to ML datasets. At a recent AAAI symposium on the UX of AI/ML, a discussion on this point led to the idea of MVD (minimum viable data (MVD), which focuses on identifying the least amount of data that will lead to a well-performing ML model. If possible, designers should work with experienced data scientists to find the MVD, taking into consideration the type of data, the desired outcomes of the ML model for the user and the implementation strategy.

ML model

Creating a useful model trained on a given data set requires expertise, time and experimentation. Often self-built machine learning models are slow, inaccurate and prone to mistakes and implementation problems.

As a workaround, it is possible to download pretrained ML models. For example, there is MobileNet (https://ai.googleblog.com/2017/06/mobilenets-open-source-models-for.html), which is optimized for mobile devices, trained on a standard data set of labelled images. There are also experts who train, build and make publicly available models for specific ML frameworks such as Caffe (https://caffe.berkeleyvision.org) or PyTorch (https://pytorch.org). These can be found in online collections such as https://modelzoo.co. Each prebuilt model may be based on different data sets and have particular performance characteristics.

Challenges with open-source data sets and models

Using these open-source data sets and models is not always an ideal solution. For example, in testing my own Delft AI Toolkit, I found that if I showed the system's vision system a current-generation smartphone, it recognizes it as an iPod. This is a result of an old image database that is the basis for the prebuilt model I used. You can see this yourself if you go to http://www.image-net.org/search?q=phone to see what outdated images are classified as phones.

Even well-funded and expert organizations like Google make mistakes in this realm. Their Photos app had a horrifying racism problem with the automatic photo tagging system. It labelled people with dark-toned skin as gorillas (Simonite, 2010) (it was still not fixed (Vincent, 2018) two years after it was first acknowledged – sadly Google worked around it by eliminating 'gorilla' as a term that can be searched). This was most likely because the system was trained on an image data set composed of primarily lighter skin-tone people but also containing a lot of gorilla images. The software mistook dark-toned skin as a characteristic of gorillas (an indicator that the product team was likely not diverse enough to notice the problem). Note that Google Photos used a system called Inception/GoogLeNet, an easily downloaded model, which is trained on ImageNet, which indeed has very few Black people in it, many of them mislabelled whites in blackface (Monea, 2019).

If you are using data sets and ML models from outside sources for prototyping, it is critical that you know their provenance and test their performance so you find what kinds of biases are lurking in them – whether they be over-represented small devices from fifteen years ago, or photos of gorillas outranking Black people.

Tools for sketching and prototyping

- **RunwayML**: ML for creators, https://runwayml.com – simple visual interface, for cloud-hosted models
- **ML5**: friendly ML for the web, https://ml5js.org – easy for quick demos and those experienced with JavaScript and web development
- **Wekinator**: http://www.wekinator.org – for artists and musicians working with sensors; communication is via the OSC protocol. The free online course is a good intro for designers/artists to ML
- **Delft AI Toolkit, tool for prototyping AI**: https://github.com/pvanallen/delft-ai-toolkit – a visual authoring environment that works in simulation mode as well as with a physical robot.
- **Teachable Machine**: a web-based tool for creating quick ML models, great for experimenting and familiarizing with ML – https://teachablemachine.withgoogle.com, https://www.youtube.com/watch?v=T2qQGqZxkD0
- **Cloud-based services**: these often require coding to interface with them but also include rich authoring systems to create the models.

- **IBM Watson**: https://www.ibm.com/watson – an integrated collection of 'cognitive services'
- **Microsoft Azure Cognitive Services**: https://azure.microsoft.com/en-us/services/cognitive-services/
- **Google AutoML**: https://cloud.google.com/automl/ – good if you want to use your own data to create a model

Sketching and prototyping collaborations

Creating smart objects involves collaboration between many disciplines. One of the key roles is the data scientist. A data scientist is someone who works with the data and algorithms that form the ML model, whether that data are ten thousand images of faces or sensor data coming from an industrial IoT enabled factory. Data scientists are often experts in several areas including collecting, cleaning and labelling data, training, testing and optimizing ML algorithms

I interviewed several practitioners and researchers in AI design, and their perspectives on designers working with data scientists are summarized here.

John Zimmerman, Tang Family Professor of AI and HCI, Carnegie Mellon University

Designers who have some basic data skills and who are given access to log data or other relevant data ahead [… of time …] have a much better experience collaborating with data scientists. In my experience, these designers can look into the data to get a felt sense if the correlations they think are happening might actually be happening. When they do this ahead of time, they look less naive when they ask about possibilities.

Having designers and data scientists envision new ideas together can be quite effective. This keeps them both committed to an idea instead of one side feeling they are working in service of the other side.

Kyle McDonald, Artist (https://kylemcdonald.net)

One of the main patterns I've seen is a push-and-pull between what a designer believes is possible and what a data scientist or developer believes is possible. Usually, they're both wrong. After an initial discussion, the designer might walk away realizing they fell for some hype around a new technology. And the developer might realize that there's a simple way to push tech in a direction it hasn't gone before.

Chris Noessel, Senior Design Lead, IBM

Data scientists are excellent partners for providing insights to what is actually happening with existing systems (and often why), as well as helping teams to ground ideas in real data. I certainly enjoy co-locating with data scientists for frequent, informal interactions and feedback.

Qian Yang, HCI researcher, Carnegie Mellon University

Doing user studies early. Sharing design ideas with them early. Talking with them more often in general, to get a general understanding of what data sets and techniques are available; what are the performance problems easy to fix versus not.

Mike Kuniavsky, R&D Senior Principal, Accenture

By identifying useful constraints for what is useful, acceptable and valuable to the human consumers of AI-generated information. For example, while 98 per cent accuracy may seem great, it may not be enough for a radiologist to trust the system completely, so positioning it as a replacement is likely not workable, but that accuracy can be useful as a assistance tool for the same radiologist. The experience will have to be designed completely differently, however.

David Young, Artist (http://www.triplecode.com)

Be willing to get your hands dirty.

Insights from practitioners on sketching and prototyping

John Zimmerman, Tang Family Professor of AI and HCI, Carnegie Mellon University

We have found that matchmaking (Bly & Churchill, 1999) is a great design technique for trying to innovate with AI. This works better than user-centred design. With matchmaking, teams start with a technical capability and then search for a customer/user that might benefit from this capability.

We have found three effective starting places for using matchmaking:

a. Start with a technical capability. We usually use a commercial service. For example, spam filters are a two-class classifier that sort documents into two piles: spam and not spam. Using this as a starting place, have teams envision as many uses as possible for a system that can sort documents into two piles.

b. Start with a data set. This is a kind of resource companies often possess. Data sets often have value for someone other than current users – user-centred design will never find this. As an example, companies in the 1990s put GPS units on trucks to improve logistics. By doing this, they inadvertently created traffic flow data. This has value they could resell to traffic services. Matchmaking with a data set has teams interrogate a data set for inferences that can be drawn, and the team searches for customers that might benefit from those inferences.

c. Start with a platform. This has teams look at the capabilities of a platform (e.g. smartphone) and investigate the sensors and data it has access to that might produce interesting applications.

Kyle McDonald, Artist (https://kylemcdonald.net)

There is abundant low-hanging fruit waiting to be exploited for short-term publicity. But to make any real impact, significant work is needed around mundane tasks like collecting and labelling data sets, training networks or evaluating different technical systems. Machine learning can be incredibly useful, but in the end there's nothing magical about it. You're still working with computational automation, and there will always be a lot of boring preparation that goes into that. In building something new, expect to spend a lot of time scraping data, cleaning it and post-processing the output of whatever algorithm you're working with.

Chris Noessel, Senior Design Lead, IBM

It depends on how the AI is manifest, what it's doing for and with the user. Simple machine learning inputs and outputs can be comped interactively with tools like InVision. More complicated AI that utilize, say, machine vision will likely require static comps in something like Sketch or Figma. Natural language interfaces are tough, requiring static examples or person-behind-the-curtain demos. For objects, I still have a soft spot for Arduino, though I haven't had my hands on one in years. There are some ideas that must be prototyped in code, and I encourage designers to learn something like processing for easy, designer-friendly programming environments.

Qian Yang, HCI researcher, Carnegie Mellon University

I have mostly created Wizard-of-Oz (WoZ) systems and interfaces for prototyping AI applications. WoZ could manifest the interaction designs realistically enough for me and my user study participants to get a felt understanding of the design's UX in particular moments. However, WoZ systems cannot realistically simulate the AI system performance change **over time** or with different user history and use contexts. This means WoZ is unlikely to catch unintended interactions AI systems sometimes make (e.g. AI biases, culturally insensitive interactions). I have not yet found a solution to this.

I think the biggest challenge of designing AI is the tension between a design workflow and a data science workflow.

Designers often ask data scientists: What are the things AI is good at? How well can the system perform? They want a definitive boundary of technical capabilities available, in order to conceptualize the design space and available technical solutions. Data scientists often ask designers 'what do users want?' They want a definitive AI learning goal in order to focus on improving system performance.

Yet for AI, there seems no definitive boundary of what AI can and cannot do; it often depends on what data users have generated and how predictive their interactions are. There are probably no AI predictions that users definitely want; how much users want the prediction depends on the quality of the prediction (e.g. accuracy). In order to design a technically achievable AI interaction that users appreciate, both designers and data scientists will need to collaborate and learn to handle these new uncertainties in their work.

Mike Kuniavsky, R&D Senior Principal, Accenture

The biggest is that there are many different kinds of AI, even within the machine learning space, and each of them has different capabilities and constraints. Using each form of AI is as much a question of finding problems that work well within those envelopes as much as adapting the technology to fit problems. Some problems are impossible for one kind of AI, but well understood for another.

Designers should not think of AI as some kind of undefinable magic dust, or only as machine/deep learning or computer vision. It's a rapidly evolving field that doesn't have a fixed set of material qualities but does have strong underlying principles that will guide its future development. Essentially, it's more a set of capabilities and potentials, a way of thinking, rather than a specific set of technologies or use cases. This means that close collaboration between data scientists/AI developers, designers and user researchers benefits everyone involved.

Conclusion

Smartness is a new and unique design material that has particular characteristics, affordances and limits, which need new design perspectives and strategies. These smartness qualities include unpredictability, unintended outcomes, animism, complex human-AI communication and social relations.

In this chapter, I have focused on sketching and prototyping smart objects, where these approaches need to take into account the special nature of smartness as a design material. This is challenging because without the assistance of data scientists and other specialists, designers may not be familiar with the capabilities of the different types of AI and ML systems. There are also specific challenges of training those systems, including the need for large, carefully cleaned and labelled data sets, avoiding bias in data collection and problems that arise from low-cost (time and money) ML models which are useful for rapid sketching and prototyping.

Meeting these challenges may require new approaches, such as using exploratory, speculative and provocative briefs to develop project concepts. Designers may also have to alter conventional human-centred design strategies because the beneficial qualities of smartness often go against UX principles such as consistency and transparency. In addition, designers may have to stretch themselves and do some technology tinkering to develop better tacit knowledge of smartness and the data that enables it.

Practitioners' experience also tells us that effective design of smartness comes from engaged collaborations with data scientists and that this collaboration is a two-way street, where designers and the data scientists are productively challenged to see their work in new ways.

Bibliography

Birhane, A., & van Dijk, J. (2020). Robot rights? Let's talk about human welfare instead. In *Proceedings of the AAAI/ACM Conference on AI, Ethics, and Society* (pp. 207–213). https://doi.org/10.1145/3375627.3375855.

Bly, S., & Churchill, E. F. (1999). Design through match-making: Technology in search of users. *Interactions*, *6*(2), 23–31.

Buxton, B. (2010). *Sketching user experiences: Getting the design right and the right design.* Burlington, MA: Morgan Kaufmann.

Buxton, B. (n.d.). What sketches (and prototypes) are and are not. Retrieved 7 March 2021 from https://www.cs.cmu.edu/afs/cs/Web/People/bam/uicourse/Buxton-SketchesPrototypes.pdf.

Carpenter, V. J. (2017, 26 October). Sketching in hardware 2017. *Medium.* Retrieved 14 May 2020 from https://medium.com/@VanessaJuliaCarpenter/sketching-in-hardware-2017-4c03fd866ff2.

Crawford, K., & Paglen, T. (2019). Excavating AI: The politics of images in machine learning training sets. Retrieved 7 March 2021 from https://excavating.ai/.

Dore, F. (2009, 2 October). Sketching in hardware is changing your life, by Fabricio Dore. *Core77.* Retrieved 8 May 2020 from https://www.core77.com/posts/14769/Sketching-in-Hardware-is-Changing-Your-Life-by-Fabricio-Dore.

Frank, T., Pührerfellner, M., von Rechbach, B., & Lechner, D. (2017). Log files: Stories from the internet of things. In *Proceedings of the Seventh International Conference on the Internet of Things – IoT '17* (pp. 1–2). https://doi.org/10.1145/3131542.3140279.

Gardner, H. (1983). *Frames of mind: The theory of multiple intelligences.* New York: Basic Books.

Gonfalonieri, A. (2018, 26 November). A beginner's guide to brain-computer interface and convolutional neural networks. *Towards Data Science.* Retrieved 14 May 2020 from

https://towardsdatascience.com/a-beginners-guide-to-brain-computer-interface-and-convolutional-neural-networks-9f35bd4af948.
Hooker, B., & van Allen, P. (2017). The internet of enlightened things: AI in the neighborhood. *Media Design Practices*. Retrieved 9 May 2020 from http://mediadesignpractices.net/research/ioet/.
The internet of enlightened things. Retrieved 7 March 2021 from https://mdp.artcenter.edu/news-event/the-internet-of-enlightened-things/.
Jobin, A., Ienca, M., & Vayena, E. (2019). The global landscape of AI ethics guidelines. *Nature Machine Intelligence, 1*(9), 389–399.
Kuniavsky, M. (2014). Sketching in hardware. *Make: Community*. Retrieved 8 May 2020 from https://makezine.com/2014/08/13/sketching-in-hardware-2/.
Lasnier, G. (2016, 26 October). Pedestrians may run rampant in a world of self-driving cars. *UC Santa Cruz*. Retrieved 7 March 2021 from https://news.ucsc.edu/2016/10/pedestrians-self-driving-cars.html.
Millard-Ball, A. (2018). Pedestrians, autonomous vehicles, and cities. *Journal of Planning Education and Research, 38*(1), 6–12.
Monea, A. (2019). Race and computer vision. *MediArXiv Preprints*. Retrieved 27 February 2021 from https://mediarxiv.org/xza9q/.
NHTSA (n.d.). Vehicle-to-vehicle communication. Retrieved 7 March 2021 from https://www.nhtsa.gov/technology-innovation/vehicle-vehicle-communication.
Prabhu, V. U., & Birhane, A. (2020). Large datasets: A pyrrhic win for computer vision? Retrieved 7 March 2021 from https://arxiv.org/pdf/2006.16923.pdf.
Redström, J. (2005). On technology as material in design. *Design Philosophy Papers, 3*(2), 39–54.
Schön, D. A. (1984). *The reflective practitioner: How professionals think in action*, Vol. 5126. New York: Basic Books.
Shopikon (2014, 8 April). 15 useless product designs. *Plain Magazine*. Retrieved 14 May 2020 from https://plainmagazine.com/15-useless-product-designs/.
Simonite, T. (2010, 1 November). When it comes to gorillas, Google Photos remains blind. *Wired*. Retrieved from https://www.wired.com/story/when-it-comes-to-gorillas-google-photos-remains-blind/.
STAMPS (n.d.). Sophia Brueckner on tech, humanities, and futures. *Stamps School of Arts & Design*. Retrieved 7 March 2021 from https://stamps.umich.edu/creative-work/stories/brueckner-tedx.
Treffert, D. A. (2014). Savant syndrome: Realities, myths and misconceptions. *Journal of Autism and Developmental Disorders 44*(3), 564–571.
Urquiza-Haas, E. G., & Kotrschal, K. (2015). The mind behind anthropomorphic thinking: Attribution of mental states to other species. *Animal Behaviour, 109*, 167–176.
van Allen, P. (2017). Reimagining the goals and methods of UX for ML/AI. In *The 2017 AAAI Spring Symposium Series: Technical Reports*. http://aaai.org/ocs/index.php/SSS/SSS17/paper/view/15338/14581.
van Allen, P. (2019, 21 November). Critical prototyping. *Medium*. Retrieved 14 May 2020 from https://medium.com/@philvanallen/critical-prototyping-8bc5054883d5.
van Allen, P., & Marenko, B. (2015). Reimagining interaction through animistic design. In Proceedings of the 4th Participatory Innovation Conference (PIN-C 2015) (pp. 492–499). Retrieved 7 March 2021 from https://www.researchgate.net/profile/Rianne-Valkenburg/publication/277006626_Reframing_Design_Proceedings_of_the_4th_Participatory_Innovation_Conference_2015_PIN-C2015/links/555d912608ae6f4dcc8c3b84/

Reframing-Design-Proceedings-of-the-4th-Participatory-Innovation-Conference-2015-PIN-C2015.pdf.
Vincent, J. (2018, 12 January). Google 'fixed' its racist algorithm by removing gorillas from its image-labeling tech. *The Verge*. Retrieved from https://www.theverge.com/2018/1/12/16882408/google-racist-gorillas-photorecognition-algorithm-ai.
Wang, P., Sibi, S., Mok, B., & Ju, W. (2017). Marionette: Enabling on-road Wizard-of-Oz autonomous driving studies. In *Proceedings of the 2017 ACM/IEEE International Conference on Human-Robot Interaction (HRI '17)* (pp. 234–243). https://doi.org/10.1145/2909824.3020256.
Yang, Q., Steinfeld, A., Rosé, C., & Zimmerman, J. (2020). Re-examining whether, why, and how human-AI interaction is uniquely difficult to design. In *Proceedings of the 2020 CHI Conference on Human Factors in Computing Systems (CHI '20)* (pp. 1–13). https://doi.org/10.1145/3313831.3376301.

8 CO-DESIGNING AND CO-SPECULATING
ON DIFFERENT FORMS OF DOMESTIC SMART THINGS

William Odom, Arne Berger and Dries De Roeck

The interaction design community has long researched *the home* and applied diverse methods to these investigations. This trajectory of work has produced important contributions that have shaped how 'smart' computational objects can be designed to better support the tasks, routines, and experiences of home life (e.g. see Desjardins, Wakkary & Odom, 2015). However, conceptualizations of what the home is, how it is made and by whom have remained somewhat narrow in the interaction design community. Whether implicitly or explicitly, 'the home' is often characterized as a detached house and 'domestic life' cast as the social organization of collocated family members (e.g. heterosexual couples with children). This critique follows a strand of literature in science and technology studies (STS) that has shown how the design of technology often reinforces existing social roles and ideas of the home (Cowan, 1983; Martin & Mohanty, 1986). These works make clear that any change in social roles that seek to challenge emergent forms of home and promote diversity and difference ought to be mirrored by changes in technology.

There is a need for new approaches to co-designing and co-speculating on emerging smart objects and Internet of Things (IoT) technologies that challenge, rather than reinforce, social roles and narrow concepts of 'the home'. This need resonates with social theorist Ursula Franklin's notion of holistic technologies that go against reinforcing "a culture of compliance" (1999, p. 24) and draw attention to the co-constitutive nature of technology:

> Technology has built the house in which we all live. The house is continually being extended and remodeled. More and more of human life takes place within its walls, so that today there is hardly any human activity that does not occur within this house. All are affected by the design of the house, by the division of its space, by the location of its doors and walls. (*ibid.*, p. 11)

For Franklin, technology is a pervasive social phenomenon that shapes our lives. Yet, in contrast to a technological determinist stance, which posits technology as a force that largely determines social phenomenon, the metaphor of a house reminds us of the critical and social roles that we, as designers and researchers, play as architects of technological systems.

A goal of this chapter is to offer a step towards expanding the interaction design community's approach to conceptualizing and designing for 'the home', domestic life and smart objects in this diverse context. This chapter describes and reflects on two design cases that offer different, yet complementary, approaches to designing domestic technology through involving a diverse set of people living in different kinds of domestic situations that exist largely outside of a 'mainstream' view of the home.

The first case focuses on *Different Homes* – a project that consisted of the use of cultural probes and design ethnography with people living in various kinds of home environments (e.g. in a boat, van, micro-loft, remote tiny house). Insights from these research activities inspired the creation of speculative design proposals that envision different ways that domestic technology could support a wider set of values, needs and desires by people living in different kinds of homes. The combined approach of conducting ongoing ethnographic work, crafting and deploying cultural probes, and generating design proposals produced new insights into and questions about how the interaction design community might conceptualize designing smart everyday things for a more diverse set of dwellers.

The second case focuses on *Loaded Dice* – a co-design toolkit that centres on the use of two 3D-printed cubes consisting of various sensors in one cube and various actuators in the other. The Loaded Dice toolkit was used to support generative activities with co-designers from a wide variety of backgrounds, abilities and domestic situations to better understand how future smart connected objects could be created to support their unique needs, values and desires. Findings from participatory workshops using the toolkit revealed that they were effective at enabling people from a wide variety of backgrounds to generate a range of idiosyncratic design concepts. The proposed design concepts offer an alternative vision of how smart objects could be designed for these specific people's everyday lives in and around the home.

Taken together, these two cases offer different accounts of how designers and researchers can approach co-designing and co-speculating with people in the service of envisioning new ways of designing smart objects. Through describing and reflecting on the benefits and limits of each approach, this chapter aims to expand the design space encompassing domestic smart objects as well as raise new questions to frame future research and practice.

Acknowledging and designing for different homes

Background

The Different Homes project is situated in the Vancouver, Canada Metropolitan area. Like many cities worldwide, Vancouver is facing numerous challenges in the areas of affordable housing and availability of space to accommodate growing population density. These issues and a range of social motivations have catalysed a growing number of citizens in the Vancouver area to adopt living situations that are smaller, mobile, temporary, self-made and/or collective. Our goals in this project are to (1) better understand the values and practices of people that embrace living situations that could be considered 'alternative' to mainstream domestic dwellings, and (2) to critically inquire into how such insights could inspire new ways of thinking about designing for 'the home' and what such a design practice might entail. Specifically, we were interested in several related questions:

- What would a 'smart home' be in the context of such alternative dwellings?
- What do connected objects mean if you frequently move between zones connectivity and disconnectivity?
- What kind of small luxuries are indulged in when there may be limited space for them?
- How do you build a record of home over time when home is not fixed to specific geographical location or set of household members?

Approach and method

Our design research inquiry across this multi-year project was divided into two stages: (1) a cultural probe (Boucher *et al.*, 2018; Oogjes, Odom & Fung, 2018) and ethnographic research (Odom, Anand, Oogjes & Shin, 2019; Shin, Sepúlveda & Odom, 2019) with people that adopted different living situations, and (2) developing speculative design proposals that responded to the values, desires, motivation, practices and experiences of our dweller participants (Oogjes *et al.*, 2018; Odom *et al.*, 2019).

For the first stage, we recruited a diverse set of participants that permanently lived in settings such as a van, boat, micro-loft, tiny house, urban condo, collective house and across many dwellings (as a house/pet sitter). To better understand their lives, we initially conducted a cultural probe study (Oogjes *et al.*, 2018). Cultural probes enabled participants to reveal to us their lives and ways of enacting domesticity on their own terms. To complement the breadth of the cultural probes study, we conducted an eight-month ethnographic study of dwellers living in three separate collective homes and dwellers living in three separate mobile dwellings (one van dweller and two boat dwellers) (see Odom *et al.*, 2019; Shin *et al.*, 2019). Mobile dwellers tend to live in vehicles where the interior of their home environment is relatively fixed, while the exterior environment surrounding their home is often changing. For collective dwellers, the physical location of the house is fixed, while the inhabitants (and objects) residing in the home may change over time. Our decision to conduct longer-term field research with collective and mobile dwellers enabled us to go deeper into understanding key overlaps and differences in their perspectives, values and ways of socially and materially organizing the home.

For the second stage, we drew on the returned probe materials and examples from our field research for design inspiration to speculatively engage with different considerations of the home and the role of technology within them. Our aim was to cultivate an attitude towards design for other, less considered forms of domestic life and to open up a dialogue about different ways that domestic technology could be explored in the interaction design community. We were particularly inspired by prior work that has focused on the creation of fictional products and product catalogues (e.g. Bleeker *et al.*, 2014). We were drawn to their capacity to catalyse a sense of familiarity at first glance through the styling of an advertisement, while then sparking critical reflection as the viewer recognizes, upon deeper inspection, a distinctly different technological future through the products and their attendant details.

We decided to embody the design proposals as various fictional products and services to think through how alternative domestic technologies might be used, designed and marketed. Our aim was to subvert and extend common tropes around domestic technologies. Our higher-level goal in developing these design proposals is to show that the proposed products do not exist in isolation but rather in relation to other services, products and systems within a sociotechnical world, and, in this, to question what this sociotechnical world might be like and for whom it might (or might not) be desirable.

Two speculative proposals: RoomiRoomba and Connectivity Clock

Next, we introduce and reflect on two speculative product concepts. These concepts aim to explore how insights from our cultural probes and field research translate into design concepts that explore how technology might be envisioned to fit in such unique contexts and to question underlying assumptions in the mainstream consumer technology marketplace. Our aim is to use these concepts as proposals to raise questions with our dweller participants on the potential role of new technologies in their lives and, through this process, co-design new concepts with them.

RoomiRoomba

The RoomiBoomba (Figure 8.1) concept takes inspiration from collective dwellers and how boundaries of personal and shared space were negotiated in the home. These social practices are tied to the identity of our collective homes and reinforce their commitments to living cooperatively. They require collective dwellers not only communicate with each other but also for them to explore and reflect on what their personal boundaries are. Our dwellers' desires to live cooperatively were strong, but the nuances of socially signalling personal, shared and collective time and space could be challenging. This proposal explores how a smart product service might play a stronger mediating role in this process. As highlighted in the Product Reviews and Questions section, we aimed to explore what positive and negative consequences might emerge from delegating this type of labour an autonomous smart object. Further, this concept raises questions about how smart home technologies could be designed to support social configurations of domestic space that are in constant flux, while the physical house itself remains a long-term fixed entity.

RoomiRoomba reimagines familiar-looking smart home products through the unique social practices and dynamic boundaries of collective homes. RoomiRoomba offers an example of how the behaviour and presence of a smart vacuum cleaner could be extended to play a direct role in mediating the frequently changing configurations of personal, shared and collective space, thus serving as an extension of close-knit values of the collective. Its presentation within an Amazon advertisement with both positive and negative reviews provokes questions around the potential benefits and consequences of such technologies: Where do boundaries of acceptability lie when we extend practices tied to the sensitive and delicate social values of a household to a semi-autonomous smart home system? To what extent should we leverage the largely unseen individual data produced by household members' daily activities as a resource for mediating the social practices of a collective household (or any household)?

Connectivity Clock

The proposal of the Connectivity Clock navigation app (Figure 8.2) was inspired by our dwellers' descriptions of moving in and out of digital connectivity – which

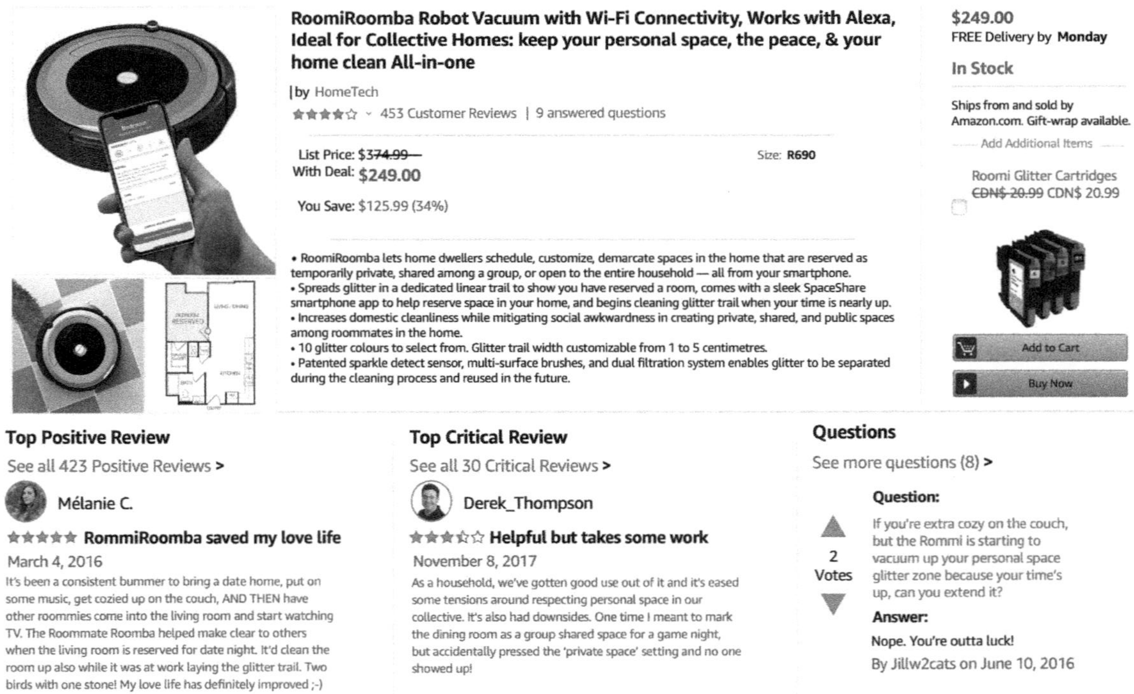

FIGURE 8.1 The RoomiRoomba is a vacuum cleaner that playfully embraces the social culture and practices of people living in collective homes.

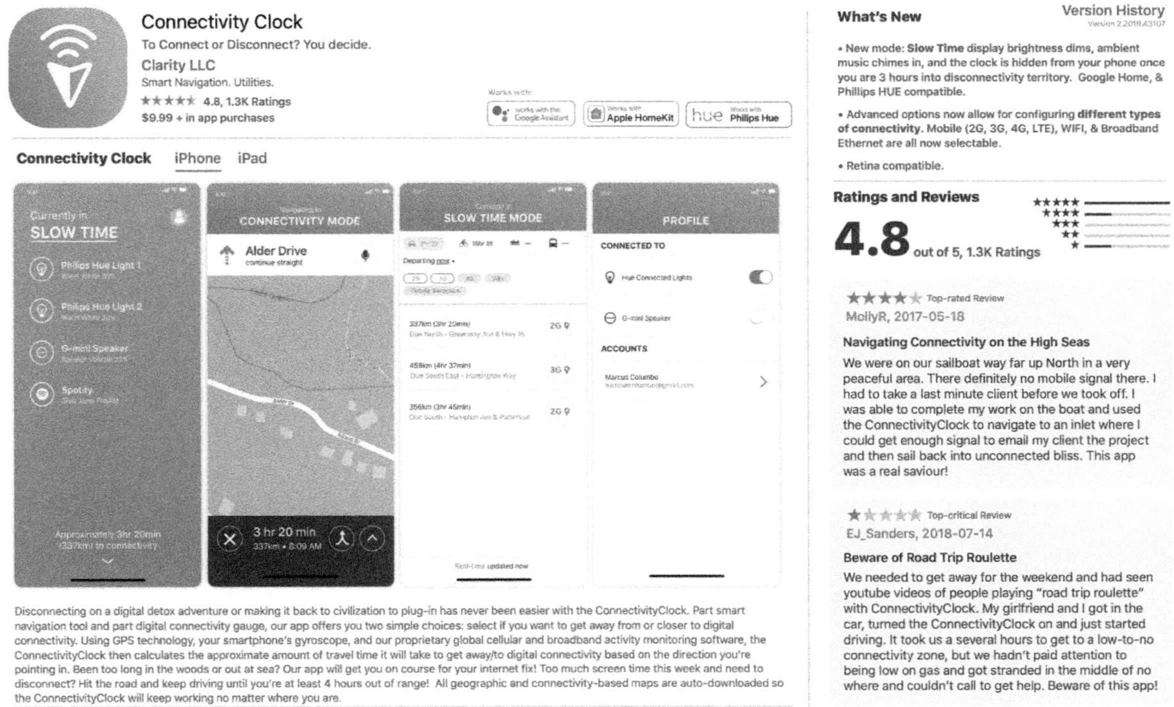

FIGURE 8.2 Connectivity Clock is a smartphone application that helps users navigate to differing levels of mobile internet (dis)connectivity.

projected 'connectivity' as a more porous, stratified and permeable concept. For example, a boat dweller mentioned how she had to sail farther north each summer to get away from smartphone connectivity. There were increasingly fewer zones that allowed her to truly get away from the connected world. Yet, she also enjoyed getting back to connectivity, the city and its infrastructure. Connectivity Clock provides information on how to direct oneself into different levels of (dis)connectivity, while not privileging one over the other. This provocation challenges the always-on ideal. Yet, it does so in a nuanced way by foregrounding freedom of choice to actively modulate one's (dis)connectivity desires across geospatial temporalities. Moreover, the Ratings and Reviews section suggests it may not be for everyone. New features such as slow time mode explore and question the desirability of enabling different levels of (dis)connectivity to open up new interactions with other locally connected devices and services (e.g. different kinds of smart light hues and music turn on once entering/leaving deep disconnectivity zones).

The Connectivity Clock proposal inquiries into the transitional qualities of our mobile dweller through a concept that leverages digital connectivity to amplify orientational awareness to changing conditions outside of the home – whether it is geographic directionality or spectrums of (dis)connectivity. Connectivity Clock recasts digital connectivity as a porous spectrum with possible richness in the stratified segments between totally connected and disconnected. This (dis)connectivity spectrum presents an intriguing space for designers to investigate in the future: How might different strengths and types of connectivity change our relation to objects, devices, people and the broader environment around us? For mobile dwellers and others alike? To what extent would this be wanted and why? The Connectivity Clock proposal open opportunities for co-design and co-speculation with dweller participants to explore how new designs could generate different kinds of geospatial awareness by considering connectivity along a wider spectrum, while still balancing people's agency and keeping them in the driver's seat.

Co-designing idiosyncratic smart objects with the Loaded Dice toolkit

Background

The Loaded Dice toolkit and workshop concept are influenced by the Scandinavian tradition of participatory design, which acknowledges that those that will be affected by a future technology ought to have an active say in its creation (Simonson & Robertson, 2012). Having people participate in designing future technology has the capacity to balance power distances between those that create and those that use technology. Designing together with people also has the potential to produce unique design outcomes that address and are aligned with the particular life worlds, values, needs and desires of the people affected by it. The mixed materiality of smart objects, however, is a particular challenge for involving people in co-design. Grappling with the complexity of intertwining tangible objects with intangible services requires a technical understanding of sensors, actuators, networks and services. It also requires having expertise in abstractly envisioning and reflecting on potential future socio-technological assemblages of objects and services, and how they might shape people, their goals and the places they inhabit.

A variety of tools and methods have been proposed to address such complexity in the design process and support people in understanding the (in)tangible components of smart objects and services. An overarching goal of these tools and methods is to empower people to become co-designers of future smart objects and services. Some such tools are the Know Cards (http://designswarm.com/portfolio/know-cards/) and IoT Design Kit (De Roeck, Tanghe, Jacoby, Moons & Slegers, 2019) which provide *abstract representations* of IoT building blocks (e.g. sensors, actuators and networks) together with contextual concepts (e.g. places, people and goals). Such tools have been shown to be highly supportive in co-designing future artefacts for a variety of contexts and settings. Yet, a known challenge is that designing with them requires some degree of abstraction (Berger, Ambe, Soro, De Roeck & Brereton, 2019a). In contrast to this, purely technical co-design tools exist. They do not include contextual features but embody functioning IoT technology, so that people can tangibly explore the functionality of networked sensors and actuators. These tools raise the challenge that people may overly focus on the technicalities of the toolkits, while having trouble focusing on contextual concepts and socio-technological connections (Ambe *et al.*, 2019).

These challenges can be addressed with situated co-design workshops that take place in people's homes (*ibid.*; Berger *et al.*, 2019a) and carefully combine co-design approaches from both realms. One such approach is the Loaded Dice toolkit that we detail below. This toolkit combines a card-based workshop to explore problem-solution spaces in individual domestic settings with a functional IoT toolkit that makes networked sensors and actuators tangible.

Approach and method

The workshops conducted with the Loaded Dice toolkit take place in people's homes. They start with co-designers exploring and explicating a particular problem, goal or situation from their domestic lifeworlds with the help of a card set. This card set represents contextual concepts of places, people and goals as well as interaction properties. Co-designers first define an interaction goal through *Goal Cards* that help them to define a domestic problem-solution space. Subsequently, cards are used to refine this through *Actor Cards* and *Space Cards*, depicting the people and places involved in the problem-solution space. Following this, co-designers define input and output characteristics through the selection of

Property Cards representing sensor and actuator states. These *Property Cards* help to detail the particular functions, emotions and aesthetics of interactions within the problem-solution space; they define the *how* of sensor-actuator interaction. Co-designers start with a basic set of these cards but can create new cards when they find people, places or properties to be missing.

Only then, when the problem-solution space has been defined, co-designers engage with Loaded Dice that embody functioning networked sensors and actuators. Loaded Dice consists of one sensor cube and one actuator cube. On each face of the sensor cube a different sensor is located, while on each face of the actuator cube a different actuator is located. Both cubes are wirelessly connected and interact with each other: The upward-turned face of the sensor cube senses and communicates sensor data to the upward-turned face of the actuator cube. Turning different faces upwards activates the corresponding sensor or actuator. The sensor cube has one of six sensors on each face: potentiometer, microphone, infrared thermometer, lux-meter, passive infrared detector and ultrasonic transceiver. The actuator cube has one of six actuators on each face: Peltier element, vibration motor, LED bar graph, fan, loudspeaker and power LED.

Loaded Dice supports co-designers in tangibly exploring the functionality of and interaction between sensors and actuators. Co-designers can tangibly explore what it means, for example, to sense heat (infrared thermometer) and actuate it as heat (Peltier element), or to transform the same heat into movement (vibration motor) or sound (loudspeaker) by simply turning one cube.

Co-design workshop strategies

Both the Card Set and Loaded Dice support co-designers to first define a goal, be it problem, need, value or dream, they want to tackle with a smart object. Co-designers then tangibly explore the possible IoT functions and interactions through repeated turns between the Card Set and Loaded Dice. Following this rationale, we conducted Loaded Dice workshops with co-designers from various backgrounds, age groups and domestic living situations, paying particular attention to people's lifeworlds. Workshops have been conducted in the homes of co-designers to encourage them to take ownership of the co-design process and to actively explore problem-solution spaces within their home.

The Loaded Dice workshops enabled haptic, associative, functional and idiosyncratic strategies of relating IoT capabilities to their individual living situations, needs and desires. Often, these strategies merged into each other and represent different stages of fluency of co-designing future smart objects. Some such future smart objects are sophisticated engineering solutions for indoor-navigation systems, emotional connections over a distance or systems to automatically feed pets or water plants. Next, we detail two ideas for smart objects that explicitly follow idiosyncratic strategies. They are idiosyncratic in the sense that they rely on particular sensible negotiations of emotional and sensory qualities and situated knowledge of lived experiences in domestic spaces. As such, they highlight how people associate unique goals with individual feelings of attachment, desires for well-being and dreams of how 'the home' is imagined. *The Whether Bird* and *The Inflatable Cat* depart from current norms in mainstream product design that focus on creating a more efficient domestic life. Instead, they shed light on how people imagine future domestic life with smart technology and relate computational smartness to their individual needs, as well as the 'sticky life situations' tied to their domestic routines.

Two idiosyncratic smart objects: The Whether Bird and The Inflatable Cat

The Whether Bird

The Whether Bird (Figure 8.3) is an idiosyncratic conceptualization of a smart object that emerged from a workshop with visually impaired students (Lefeuvre *et al.*, 2016). The student co-designers disapproved of speech assistants because using them might expose the user as 'needy and handicapped'. Simultaneously, they face the problem that their smartphone apps only provide weather forecasts with no way of knowing whether it had rained and the streets would still be wet:

Researcher:	"How do you know if it did rain overnight?"
P04:	"I ask a Weather App since I can't look out of the window. Otherwise I would notice when I feel that the street is wet."
Researcher:	"Don't you are at risk getting wet feet then?"
P03:	"Been there."
P04:	"You also can smell whether it did rain."
P03:	"Right!"
P02:	"I feel like the birds sing more melancholically when rain is approaching."

FIGURE 8.3 The Whether Bird sings more melancholically when rain is approaching.

In answering these two challenges, students envisioned The Whether Bird which we describe in a short scenario: Outside, on the windowsill, a weather sensor would measure the amount of rain over the past few hours. Inside, within the flat, a plush bird equipped with a hidden actuator would be wirelessly connected to the weather sensor. The plush bird would sing at the touch of a button, a tweak to the beak, or by stroking the birds' belly. Depending on whether it has rained, the bird would sing just a slight bit differently, so that only the blind student would know what this means.

The Loaded Dice workshop empowered the student co-designers to explore an issue from their domestic realm and to ideate a blueprint for future smart object. With the Loaded Dice toolkit, student co-designers did not just combine merely functional IoT building blocks to a smart object. Instead, they engaged in an immersive sensory-oriented exploration of goal and context, while simultaneously outlining the technical details of a future smart thing. The students co-designed a smart object that would not focus on deficits but instead foregrounds the extraordinary perception of blind people.

Many designs for people living with blindness stem from a deficit-based approach where assistive technology is engineered to make blind people better fit into the routines and capabilities of an able-bodied world. Our co-design approach enabled blind co-designers to voice their desire for technology that does not stigmatize them. It enabled blind people to ideate and then propose a technology that solves a problem form their life world by focusing on their innate abilities. The Whether Bird illustrates how co-designing artefacts with those that will be affected by them can lead to designs that support individual desires and capabilities. The Whether Bird also actively questions the normative, efficiency-oriented narrative of mainstream smart home technologies and advocates for a more situated, bottom-up approach to smart object design.

The Inflatable Cat

The Inflatable Cat (Figure 8.4) is a vivid example of how people co-designed a smart object for a 'sticky life situation' that they do not know how to solve well. It involves their cat within the context of their communal living arrangement (Berger *et al.*, 2019b). The aim of the smart object is to support the cat in 'what he actually desires'. Also, the communal house where the co-designers dwell has no cat flap. This has led to the cat being out in the cold and meowing in front of closed doors in hopes he would be let in. The co-designers envisioned a concept that would enable the cat to grab the attention of the communal dwellers. The smart object consists of a microphone outside the front door that could recognize the meows of the cat and distinguishes it from other cats to raise attention in the flat. Out of the several attention-grabbing and poetic ideas co-designers articulated, their final idea is particularly idiosyncratic: Within the communal home, a fan, instead of a

FIGURE 8.4 The Inflatable Cat supports a real cat in what he actually desires.

loudspeaker, would actuate the presence of the cat at the front door. To not disturb conversations, an oversized balloon-like version of the cat would be inflated by a fan, which would subsequently rise to the ceiling and vibrate.

The Inflatable Cat, as a co-design outcome, illustrates how the participatory workshop setup encouraged co-designers to reflect on poetic aspects of distributed communication while associating suitable sensors and actuators embodied by Loaded Dice. The Loaded Dice toolkit workshops enabled co-designers to explore problem-solution spaces closely aligned to their domestic experiences and to imagine smart objects that individually fit these contexts. The creative strategies exhibited here are idiosyncratic in the sense that the co-designers created ideas for smart objects that are specifically make sense within their housing situation. More generally, these co-design workshops and outcomes can help to better understand what people call 'the home' and what people consider to fit well into their very own, individually situated situation and socially constructed boundaries.

Discussion and conclusions

Designing interactive systems intended to support people's everyday lives and practices at home continues to raise opportunities and issues for the interaction design community. Design has long been regarded as an approach for framing, setting and solving human problems, and improving the conditions of people's everyday lives. Yet, design can also operate as an approach for critically provoking, imagining and questioning how we might treat such complex notions as 'the home' and the technologies designed in relation to it. A goal of this chapter has been to extend prior research by taking a step towards describing and unpacking two approaches to co-designing and co-speculating on the roles new kinds of technologies could play in different kinds of homes. These two approaches are different, yet complementary. They are both bottom-up approaches in the sense that they enable potential future users to co-speculate and co-design possible futures with smart objects. In this way, they enable a conversation between designers and future users that connects present-day domestic life with individual potential futures. Both approaches illustrate alternative ways that technology could be designed for the home, embody different ideas of where home is located, explore how home is constructed, remade, curated and pursued, and question material, technological and social boundaries between it and the outside world. They generate knowledge about alternative ways of conceptualizing future smart objects in ways that often go unseen in commercial one-size-fits-all approaches to designing smart home technology.

These approaches differ in how they involve stakeholders and how they articulate the emerging design proposals. The Different Homes project combines empirical, inspirational and speculative approaches to challenge and expand what 'the home' is and whom the dwellers are that ought to be considered by designers. The Loaded Dice toolkit workshops offer examples of how a time-constrained co-design workshop method can lead to outcomes that open up people's imaginations of future smart technology design based on their own unique idiosyncratic values, desires and practices. Importantly, across the two projects, our aim is to not be prescriptive or conclusive. In line with this book's broader goal of establishing a research program for interaction design, our goal is to raise new questions that inspire, frame and expand future research in ways that move beyond narrow assumptions of a one-solution-fits-all approach for smart home design. Through our collective inquiry across the Different Homes and Loaded Dice projects, new questions have emerged that can be circumscribed into the following key areas and serve to guide future research.

Diversifying the home

How can the interaction design community better understand and acknowledge blind spots and implicit assumptions in design research and practice? How should we better recognize factors such as geographic location, gender, race, ableism, techno-solutionism and the unquestioned commitment to scaling up technology design? And in what ways can these factors be critically engaged with through design?

These questions provoke a number of considerations for diversifying smart object design through actively engaging with people into designing, questioning and rethinking possible futures. They generate openings in the design space to take seriously the need to design for difference and the obligations that come with co-creative and co-speculative approaches. There is a need for future research to recognize and embrace more different and diverse domestic living conditions, dwellings and dwellers. This will mean engaging with people from disadvantaged communities or geographic locations that are oftentimes overlooked by the interaction design community. A starting point for these efforts will need to come with acknowledging that most co-designer and co-speculator participants, in research to date, come from a position in which they were able to choose to adopt the lifestyles they desire. Members of populations and communities that are affected by poverty, homelessness, physical/mental illness, discrimination and/or cultural annihilation (among other things) may have little choice other than to live in non-mainstream domestic conditions. Engaging with such populations represents crucially important opportunities for future research if we are to take seriously a broader, more inclusive call for diversifying the domestic and crafting new agendas for designing for a plurality of living situations.

Making it work

How can the interaction design community provide productive counter-narratives to normative assumptions of what the home is and what it entails? And how can design meaningfully 'scale-up' situated and diverse approaches to make them more available and inclusive?

The exemplars from the Loaded Dice and Different Homes projects provide a critical lens on how technology design can align with the ways that people envision their future domestic life. While these design exemplars rely on the unique values, desires, aspirations and practices of individual people and collective communities, they are not necessarily meant to be mass-produced for future use. For example, it is possible that relatively few homes might need an Inflatable Cat or RoomiRoomba. These exemplars work to concretely demonstrate that it is possible, and, in fact, sensible, to seriously engage the creative thinking of people and to trust their fluency in understanding their own domestic life. These approaches can be useful for 'making it work' in several number of ways. First, they question the ways we think about the smart home and also which 'smartness' aligns with the desires, goals and dreams of people. Second, they demand responses to the questions of what kind of 'smartness' we want and where we want it to be situated. In this way, creating concepts that might be far-fetched, critical, seemingly outrageous or even humorous help us collectively imagine and reflect on what kind of futures people want. Important in this is to take these concepts seriously enough in order to use them as design inspirations. This is well illustrated in work by Elisa Giaccardi and her colleagues on taking a thing-centred design perspective to critically question design decisions (Giaccardi, Speed, Cila & Caldwell, 2016). This work is in parallel to a growing interest in adopting and connecting co-speculative and participatory approaches in the interaction design community to engage individuals and communities in envisioning potential futures with technology and questioning if it is what they want (e.g. see Lyckvi, Roto, Buie & Wu, 2018; Desjardins, Key, Biggs & Aschenbeck, 2019).

Yet, the question remains as to what it would *mean* to make the ideas, emerging from such bottom-up approaches, work. We would need to envision new infrastructures and services to expand the notion of co-design and co-speculation to co-constructing and co-maintaining. In order to actually build the individual solutions co-designed with people, such new infrastructures would need the capacity to safely and efficiently produce and maintain smart objects as individual units or small batches.

Safeguarding and designing for the future home

How should the interaction design community ensure that the alternative futures envisioned in bottom-up approaches offer value to the people involved in their co-design and co-speculation? Is there a risk for the idiosyncratic outcomes resulting from co-design and co-speculation activities being co-opted into mainstream normative design? What kinds of unintended consequences could result? To what extent should we develop strategies for resisting against normative design?

The questions posed above highlight complex issues the interaction design community will have to face in future research. They prompt us to reflect on what kinds of 'problem-solving' we address with co-design and co-speculative approaches. They also are cautionary and make clear the need to critically consider who will benefit from the design, implementation and dissemination of such individually situated smart objects. In related research on IoT and do-it-yourself (DiY), the value of acknowledging different skills and different engagements in projects is important in order for a community to take ownership during a design process (e.g. (De Roeck *et al.*, 2012; Woo and Lim, 2015). For example, not all people are skilled at identifying valuable ways to use technology in their homes. As such, different methods and approaches are needed to critically reflect on the design proposals originating from co-design and co-speculation, to understand their effects on privacy, security and agency.

Acknowledgements

The Different Homes project acknowledges Doenja Oogjes, Sumeet Anand, Jo Shin, Peter Fung and Gabriela Aceves-Sepulveda for their important contributions to this project and the various publications and activities that encompass it. We also acknowledge that this research took place on the unceded traditional territories of the Coast Salish peoples of the Katzie, Kwantlen, Kwikwetlem (kwikwəƛəm), Qayqayt, Musqueam (xwməθkwəyəm), and numerous Stó:lō Nations. We thank our participants for generously sharing their experiences with us. This project is supported in part by the Social Sciences and Humanities Research Council of

Canada (SSHRC) and the Canada Foundation for Innovation (CFI). The Loaded Dice was made possible by Albrecht Kurze, Andreas Bischof, Sören Totzauer, Michael Storz, Teresa Denefleh, Mira Freiermuth and, most importantly, Kevin Lefeuvre. We are deeply thankful for the magnitude of support we received from Maximilian Eibl and we thank our co-designers for working with us. This project is funded by the German Ministry of Education and Research (BMBF), grant number FKZ 16SV7116.

Bibliography

Ambe, A. H., Brereton, M., Soro, A., Chai, M. Z., Buys, L., & Roe, P. (2019). Older people inventing their personal internet of things with the IoT un-kit experience. In *Proceedings of the 2019 CHI Conference on Human Factors in Computing Systems*, Paper 322 (pp. 1–15). New York: ACM.

Bettencourt, L. M. A., Lobo, J., Helbing, D., Kühnert, C., & West, G. B. (2007). Growth, innovation, scaling, and the pace of life in cities. *Proceedings of the National Academy of Sciences, 104*(17), 7301–7306.

Berger, A., Ambe, A. H., Soro, A., De Roeck, D.,& Brereton, M. (2019a). The stories people tell about the home through IoT toolkits. In *Proceedings of the 2019 on Designing Interactive Systems Conference* (DIS '19) (pp. 7–19). New York: ACM.

Berger, A., Odom, W., Storz, M., Bischof, A., Kurze, A., & Hornecker, E. (2019b). The inflatable cat: Idiosyncratic ideation of smart objects for the home. In *Proceedings of the 2019 CHI Conference on Human Factors in Computing Systems* (CHI '19), Paper 401 (pp. 1–12). New York: ACM.

Bleeker, J., Nova, N., Girardin, F., Foster, N., Byrne, E., & Tesone, L. (2014). *TBD Catalog, 9*(24). Sierre, Valais: Near Future Laboratory.

Boucher, A., Brown, D., Ovalle, L., Sheen, A., Vanis, M., Odom, W., *et al.* (2018). TaskCam: Designing and testing an open tool for cultural probes studies. In *Proceedings of the 2018 CHI Conference on Human Factors in Computing Systems* (CHI '18), Paper 71 (pp. 1–12). New York: ACM.

Campagna, G. (2016). Linking crowding, housing inadequacy, and perceived housing stress. *Journal of Environmental Psychology, 45*, 252–266.

Cowan, R. S. (1983). *More work for mother*. New York: Basic Books.

De Roeck, D., Slegers, K., Criel, J., Godon, M., Claeys, L., Kilpi, K., *et al.* (2012). I would DiYSE for it! A manifesto for do-it-yourself internet-of-things creation. In *Proceedings of the 7th Nordic Conference on Human-Computer Interaction: Making Sense Through Design* (NordiCHI '12) (pp. 170–179). New York: ACM.

De Roeck, D., Tanghe, J., Jacoby, A., Moons, I., & Slegers, K. (2019). Ideas of things: The IOT design kit. In *Companion Publication of the 2019 on Designing Interactive Systems Conference 2019 Companion* (DIS '19 Companion) (pp. 159–163). New York: ACM.

Desjardins, A., Wakkary, R., & Odom, W. (2015). Investigating genres and perspectives in HCI research on the home. In *Proceedings of the 33rd Annual ACM Conference on Human Factors in Computing Systems* (CHI '15) (pp. 3073–3082). New York: ACM.

Desjardins, A., Key, C., Biggs, H. R., & Aschenbeck, K. (2019). Bespoke booklets: A method for situated co-speculation. In *Proceedings of the 2019 on Designing Interactive Systems Conference* (DIS '19) (pp. 697–709). New York: ACM.

Franklin, U. (1999). *The real world of technology*. Toronto, ON: House of Anansi.

Giaccardi, E., Speed, C., Cila, N., & Caldwell, M. (2016). Things as co-ethnographers: Implications of a thing perspective for design and anthropology. In R. C. Smith, K. T. Vangkilde, M. G. Kjaersgaard, T. Otto, J. Halse & T. Binder (Eds), *Design Anthropological Futures* (pp. 235–248). Oxford: Routledge.
Lefeuvre, K., Totzauer, S., Bischof, A., Kurze, A., Storz, M., Ullmann, L., *et al.* (2016). Loaded dice: Exploring the design space of connected devices with blind and visually impaired people. In *Proceedings of the 9th Nordic Conference on Human-Computer Interaction* , NordiCHI '16), Article 31 (pp. 1–10). New York: ACM.
Lyckvi, S., Roto, V., Buie, E., & Wu, Y. (2018). The role of design fiction in participatory design processes. In *Proceedings of the 10th Nordic Conference on Human-Computer Interaction* (NordiCHI '18) (pp. 976–979). New York: ACM.
Martin, B., and Mohanty, C. T. (1986). Feminist politics: What's home got to do with it? In T. De Lauretis (Ed.), *Feminist studies/critical studies* (pp. 191–212). London: Palgrave Macmillan.
Odom, W., Anand, S., Oogjes, D., & Shin, J. (2019). Diversifying the domestic: A design inquiry into collective and mobile living. In *Proceedings of the 2019 on Designing Interactive Systems Conference* (DIS '19) (pp. 1377–1390). New York: ACM.
Oogjes, D., Odom, W., & Fung, P. (2018). Designing for an other home: Expanding and speculating on different forms of domestic life. In *Proceedings of the 2018 Designing Interactive Systems Conference* (DIS '18) (pp. 313–326). New York: ACM.
Shin, J., Sepúlveda, G. A., & Odom, W. (2019). "Collective wisdom' inquiring into collective homes as a site for HCI design. In *Proceedings of the 2019 CHI Conference on Human Factors in Computing Systems* (CHI '19), Paper 316 (pp. 1–14). New York: ACM.
Simonsen, J., & Robertson, T. (Eds). (2012). *Routledge international handbook of participatory design*. Oxford: Routledge.
Woo, J-b., & Lim, Y-k. (2015). User experience in do-it-yourself-style smart homes. In *Proceedings of the 2015 ACM International Joint Conference on Pervasive and Ubiquitous Computing* (UbiComp '15) (pp. 779–790). New York: ACM.

PART FOUR

CRITICAL UNDERSTANDINGS

9 MARX IN THE SMART LIVING ROOM: WHAT WOULD A MARX-ORIENTED APPROACH TO SMART OBJECTS BE LIKE?

Betti Marenko and Pim Haselager

Introduction

When talking about smart objects, the emphasis is often on the user, the experience of interaction, the design of the interface, the UX and so forth. What is not equally addressed is how the system of production of smart/intelligent objects and the organization of the forces of production behind them impacts on the resulting smart experience and environment. A Marxist critique may prove both necessary and desirable to unpack the techno-deterministic narratives of the smart home and, broadly, to recognize that the digitalization of human experience must be addressed as a political issue.

So we ask: what would a Marx-oriented approach to smart objects be like? As an exercise in philosophical thinking – almost 'philoso-fiction' – this chapter has the purpose of mobilizing critical thinking in order to reveal some insights otherwise not accessible. We begin by imagining a smart domestic environment – picture, for instance, the living room of a Silicon Valley technocrat populated by connected objects (Nest, Echo, Roomba and the like). In this perfectly plausible domestic landscape we find Karl Marx himself, sitting on the sofa taking notes on his surroundings, exactly as he did in his analysis of the Industrial Revolution.

What would Marx think? We imagine that the smart living room would appear to him as an ecosystem of alienation-inducing commodities. From this standpoint, this chapter addresses three main issues. First, the issue of alienation: It can be argued that there is a disjuncture inherent to smart objects. They possess a Janus-like quality. On the one hand they support and enable us (the user), and on the other they capture our data, time and attention. In short, they exploit us. While smart environments claim to personalize and tailor their presence to individual needs, at the same time they intrude, monitor and control our life, thus enabling new techno-digital forms of alienation where user, content provider and product all collapse in one single 'datified' dance. These new modes of alienation must be investigated. Second, how do we interpret fetishism of commodities in relation to smart objects? Marx's commodity – a mysterious thing 'endowed with life' – describes uncannily smart objects, that is, animated devices with agency. What is needed, then, is an analysis of the systems of (digital) production in order to demystify the 'necromancy' that surrounds the (digital) product of labour. When users become slaves of the very machines they created, we need design scenarios that awaken their users from the alienating slumber induced by this new opium of the people.

Digital Marx

There are several reasons why Marx is so relevant now, not least the fact that, unshackled by the historical connection to state socialism, his work can be reappraised in its luminous dark futurism.[1] It is the argument put forward by the present chapter that some of Marx's insights can assist us in unpacking the 'rebirth of the commodity-form in the triumphant language of the digital commodity-form' (Kroker, 2004, p 121). From the viewpoint of commodity fetishism, all commodities appear as hyper-sensual objects endowed with the supernatural, mystical, almost occult power to create value (Marx, 1981).

This framework seems particularly appropriate to read the smart objects that populate our environment as digital commodities, enveloped by the very same fetishism that Marx predicted. Likewise, Marx's seminal concept of alienation – the fundamental estrangement of the labourer from their labour, from society and from their own human self – seems especially pertinent to describe the contemporary ecologies of cohabitation of human and smart objects, where the role of user and worker increasingly collapse into the same individual.

Marx scholar Christian Fuchs (2017, 2018, 2019), a vocal proponent of the need to reread Marx in relation to contemporary digital capitalism, argues that Marx's notion of the general intellect 'anticipated the emergence of what some today term informational capitalism or digital capitalism or cognitive capitalism'

[1] For Arthur Kroker, Marx's intuition of the contemporary hyper-realization of the market – is the 'dark future of Capital' (Kroker, 2004).

(Fuchs, 2018, p. 526). He laments, however, the overall lack of appetite for Marxist theory in the field of media, communication and digital technologies.

The present chapter wants therefore to offer a modest contribution in this sense and, more than filling a gap, offering some insights from the unconventional perspective of a philosopher of design (Marenko) and a cognitive scientist (Haselager) using Marx to reflect on smart objects.

In the context of capitalism in its current mode (whether we call it platform, Big Data, digital capitalism, etc.), 'digital commodities' refer to the commodification of digital labour power, for instance, digital content production, upload and sharing (Fuchs, 2019).[2] While the nature of digital commodities is increasingly transparent (Negri, 2017), the character of digital capitalism is fundamentally and historically extractive insofar as it mines users' time and attention to turn them into (digital) value. Life itself becomes value-producing digital labour. The extractive character of contemporary capitalism is evident 'not only when the operations of capital plunder the materiality of the earth and biosphere, but also when they encounter and draw upon forms and practices of human cooperation and sociality that are external to them' (Mezzadra & Neilson, 2017, p. 188).

Take the model of corporate commercial social media, which demands that users create and cultivate social relations alongside an incessant production of content, all the while relentlessly mined for valuable data feeding into targeted advertising. It is this 'free' affective, digital, biopolitical labour that supports the unprecedented gigantic advertising enterprises called Facebook and Google.

The datafication of time, attention and of the whole sphere of living is a new form of capitalist accumulation.[3] This colonization of human spaces (both interior and social space) is an intrusive process. Take also the way in which a selfless and altruistic activity like 'sharing' has been appropriated and devalued, its meaning profoundly distorted by the rhetoric of social media. Human generosity, kindness and social cooperation have been emptied of their significance and turned into vectors of accumulation of value. This is another aspect of the excavating, extractive and ultimately violent nature of digital capitalism whose aim is the colonization of social relations by digital technology.

As we shall see, the connection between this extractive model and commodity fetishism's faculty to render invisible the social relations of production is evident. Again, take Facebook. While its commodity status may not be immediately clear

---

[2] Fuchs points out the non-rivalrous nature of information, which, as a resource, is not used up when consumed. On the contrary, ease of access, copying, reproduction and sharing indicate that information although highly commodified can also resist commodification. This is the contradiction existing between digital capital and the digital commons.

[3] See here Dallas Smythe's seminal work, who already in 1977 wrote that 'material reality under monopoly capitalism is that all non-sleeping time of most of the population is work time' (Smythe, 1977, p. 3; cited in Fuchs, 2019, p. 61).

as no one pays for access to its platformed sociality, it is precisely this enforced sociality that conceals Facebook's commodity form. Thus, within the sphere of social media, commodity fetishism is inverted. While in its conventional sense commodity fetishism indicates that things (commodities, money) obscure the social relations that have produced them, within corporate social media, social relations become the 'real' experience, possessing immediacy, concreteness and, most important, tradeable value (Fuchs, 2017). In other words, they become a commodity.

Alienation

Marx saw alienation ('Selbstentfremdung') as the consequence of an objectification and appropriation of labour, which stops being a self-expression but becomes a commodity (McLellan, 2000, p. 86). This applies not only to the product of labour but also to the labour itself: 'Labour is ... not voluntary but compulsory, forced labour. It is therefore not the satisfaction of a need but only a means to satisfy needs outside itself' (*ibid.*, p. 88). The labourer 'does not belong to himself in his labour but to someone else' (*ibid.*, p. 89). Workers get governed by social needs that are alien to them, and 'the greater and more elaborate appears the power of society inside the private property relationship, the more egoistic, antisocial, and alienated from his own essence becomes man' (*ibid.*, p. 128).

Smart objects run the risk of transferring this type of alienation from the work context to the home. Increasingly, the personal life of individuals and families gets orchestrated via virtual assistants or smart speakers like Alexa, AliGenie, Siri or Google home. Smart objects change our home situation from something essentially private into a product to be measured, registered, catalogued, transmitted and exploited. It is well known that some robot vacuum cleaners also function, commercially speaking perhaps even more importantly, as measuring devices of the space they clean, transmitting information about size and location of rooms and possibly even large objects within them (see e.g. Astor, 2017). Such smart objects transform one's living space into a commodity.

The interactions of an individual within the home environment, both with objects and other agents, are registered, collected, analysed, mediated and (at least potentially) directed or controlled by technology. The reasons underlying the willingness to become a data subject at home, to be datafied in one's personal life, are often connected with convenience, ease of use, simplification or externalization of daily tasks and so forth. We suggest that convenience can be seen as a (new) form of wage. In exchange for activities becoming 'easier', one hands over one's private data. But the interaction with smart objects is not a form of self-expression: It is governed by the requirements and opportunities that these artificial agents bring with them. As a consequence, no longer only one's labour but also one's private social life becomes alienated. The effects of 'home alienation' through smart objects may leave many traces. It can lead to agency confusions: is the smart speaker an agent? A person? Is it like us, are we like it? Parents report (personal communication) their young children asking, for example, Siri, 'What are you?' Recently, the family of the creator and screenwriter of the TV series 'Black Mirror' decided to remove Alexa after the son addressed his father as 'Alexa' (Tucker, 2019).

Smart objects can lead to more indirect forms of communication. They can also produce altered forms of interaction, like monitoring. The structure of the interaction can get regimented, subtly or less so, through smart objects. An important commercial aspect of smart speakers is to facilitate and stimulate consumption; it is, for instance, highly likely that family communication mediated by smart objects will be nudged towards consumption. Individually, certain capacities may get externalized to such an extent that one's behavioural repertoire, that which one is capable of, may get impoverished significantly or even beyond repair. Anecdotal reports of children interacting with, for example, Siri or Alexa indicate that at some point children ask them to do parts of their homework, for instance, by answering questions about calculations.

But ultimately most fundamental is that the experience of being-at-home will get mediated, transferred into a being-at-home-with-Alexa. This may alienate oneself from the embodied embedded flow of existence in private. Kreitmair, Cho and Magnus (2017) discuss this effect in relation to wearable and mobile health technology, and suggest that

> technology may negatively affect one's ability to 'be in the moment'. Experiencing the world and the self in a present, in-the-moment fashion, characteristic of phenomena such as flow, is associated with greater well-being. This requires first-person, introspective means of acquiring self-knowledge. Offloading the monitoring of one's mental and physiological processes onto external technologies is antipodal to such authentic experiencing.

Moreover, they suggest, there is concern that tracking and focusing on external means of gaining self-knowledge may be counterproductive to experiencing phenomena such as 'flow' and 'being-in-the-moment', which may contribute to alienation from embodied and embedded living (Kreitmair, Cho & Magnus, 2017). The experience, rather than lived, becomes represented and gets processed as such. A similar process occurs in extreme forms of self-quantification, where one's numbers, registered via fitness and food trackers, become more important than how one feels. 'It's a different sort of experience, in that the user is not engaging in an authentic way with reality' (Kreitmair, 2018). It is perhaps worthwhile here to point again to the quote from Marx given above: the greater the power of society inside the private property relationship, the more egoistic, antisocial and alienated from their own essence human beings become (McLellan, 2000, p. 128).

Kate Crawford and Vladan Joler (2018) present a remarkable patent owned by Amazon: Hidden among the thousands of other publicly available patents owned by Amazon, U.S. patent number 9,280,157 represents an extraordinary illustration of worker alienation, a stark moment in the relationship between humans and machines (Wurman *et al.*, 2016). It depicts a metal cage intended for the worker, equipped with different cybernetic add-ons, that can be moved through a warehouse by the same motorized system that shifts shelves filled with merchandise. Here, the worker becomes a part of a machinic ballet, held upright in a cage which dictates and constrains their movement." The patent (https://patents.google.com/patent/US20150066283A1/en) contains in Figure 9.1.

We cannot think of any better illustration of Marx's notion of alienation applied to smart objects. Smart objects measure, record, analyse, influence, nudge, steer and direct us. We will live our private lives and go through our intimate social interactions as being carried, measured and restricted by an invisible but effective digital cage in our own homes.

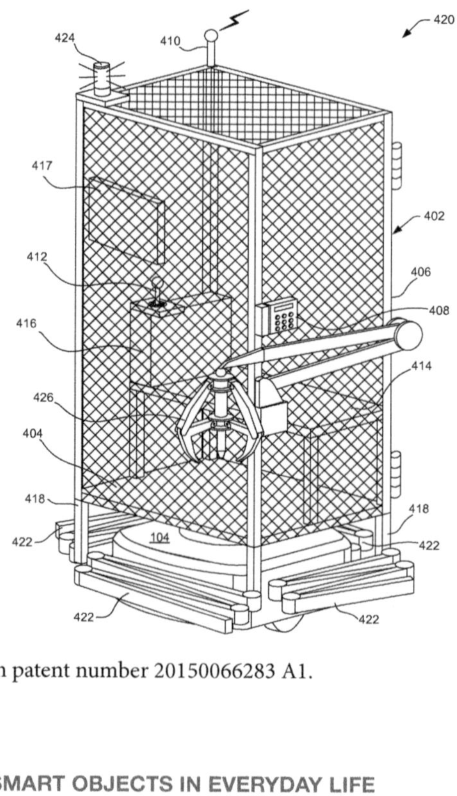

FIGURE 9.1 Amazon patent number 20150066283 A1.

Marx's commodity fetishism

In Marx's commodity fetishism, the value of a commodity embodies and hides the human labour that has created it. The commodity assumes 'the fantastic form of a relation between things' (Marx, 1981, p. 165). Thus, commodities appear to have simply emerged as independently animated, sensuous things populating the world with a definite social character. Commodities become social entities, as such able to enter into social relations not only with humans but also among themselves, beyond the world of humans. For Marx, commodities become free agents that relate 'socially' to each other but retain independence from their producers, effectively building a mystical world in which they appear to be alive, uniquely mobilized by desire.

Marx appropriated the term 'fetishism' at an early stage of his career, at a time when the literature on fetishism and the concept itself were relatively new, and pushed it beyond the confines of its received meaning. In an article dated 1842, the young Marx describes fetishism as 'the religion of sensuous desire' (Marx & Engels, 2012, p. 22), as if mocking the European ruling classes as perverse idolaters and worshippers of inanimate things.[4] The emphasis on 'sensuous desire' is crucial to an understanding of Marx's concept of fetishism and its legacy, as it foregrounds the affective dynamics of human desire seeking immediate gratification through material objects.

In Charles de Brosses's original formulation (1760; see Pietz, 1993, p. 138 n. 56), fetishism is a primitive form of religion based on the worship of artefacts, a way of thinking that attributes supernatural reasons to contingent facts and sees divine powers in terrestrial entities. Sixteenth-century European traders and colonizers would use the word 'fetish' to describe those artefacts that local tribes from the western coast of Africa would not trade as, allegedly, objects of worship. From this (Eurocentric, positivistic) perspective, fetishism was 'the pure condition of un-enlightenment' (Pietz, 1993, p. 136). On the other hand, fetishism may be interpreted as the projection of European colonial mindset and fears (McNally, 2011). What the colonizers found incomprehensible was the irreducibility of African artefacts to a price: their incalculability. Without a price there was no exchange value: These objects were out of bound, literally un-tradable. Such a refusal to trade, revelatory of the historicity and variability of market laws, had to be reframed as a perversion – a 'fetish'. Without this manoeuvre, capitalism would have had to be acknowledged as an historical phenomenon, a contingent, rather than natural event, therefore not as unavoidable as it is taken to be.[5]

---

[4] For Marcel Mauss, the notion of fetishism contains an 'immense misunderstanding between two cultures, the African and the European' (cited in Pietz, 1993, p. 133 n. 42).

[5] One may wonder what Donald Trump's attempt to buy Greenland says about the convergence, at least in the US president's mind, of capitalism with the fetishism of a specific kind of commodity: real estate.

Indeed, the key point in Marx's theory of commodity fetishism is that the 'materiality of "value" is not physical but social' (Pietz, 1993, p. 145). Commodity fetishism is ultimately an inversion process, namely the inversion of the relations between humans and things: As things are personified, people are objectified. Marx's genius was in understanding that capitalism is about relations (and their inversion) rather than material things, and that at the core of this process lies the enigma of fetishism.

This is exactly where Marx's actuality lies. He understood that capitalism is not about things, but about relations. Moreover, he predicted with uncanny accuracy the double shift, from production to consumption first, and then on to capitalism itself as the pure vector of circulation of technology (Kroker, 2004) – what we will call throughout digital, platform, Big Data or surveillance capitalism.[6]

The fetishism of technology

Marx, the great futurologist, recognized that fetishism is beyond things in themselves and resides in the fundamental immateriality of value. This has become strikingly evident more so in digital capitalism where everything (time, attention, experiences, life itself) is commodified and value is fetishized precisely because they (time, attention, experiences, life itself) are immaterial.

Marx's fetishism of the commodity can be reappraised to defamiliarize our understanding of technology and to therefore grasp differently modern technological objects with their apparent autonomy and agency (Hornborg, 2012, 2014). While money, commodities and machines are all equally fetishes as 'they mystify unequal relations of exchange by being attributed autonomous agency or productivity' (Hornborg, 2014, p. 121), what distinguishes machines is the extent to which they are often conceptually framed by a techno-determinist narrative, that is, the idea of technology as an inevitable progress over time. Put differently, technological innovation is perpetually disguising the conditions of its production, which is why technological development is presented as a natural progress rather than as capitalist accumulation, inequality and exploitation. This teleological view obscures the fact that any 'rationale of mechanization is inextricably intertwined with global differences in the price of labor and resources' (*ibid.*, p. 122). Far from being a natural progression in innovation, technology (every technology in fact) is 'contingent on specific global constellations of asymmetric resource flows and power relations' (*ibid.*, p. 122). It can therefore be argued that the 'fetishism of technology represents a specific mode of mystifying unequal exchange' (*ibid.*, p. 134) based on the denial of appropriation and on novel processes of

---

[6] Kroker uses many terms to describe late capitalism: virtual, speed, digital, hyper, streamed capitalism.

accumulation which have shifted from extracting value from physical labour into the new extractive territories of time, attention and life as a whole.

Now, if we take this perspective to look at a few examples of smart objects, for instance, digital home helpers such as Nest, Echo or Roomba, what emerges is the extent to which these devices signal a new, hypertrophic mode of commodity fetishism where the key aspects delineated above take on digital specificity. To start with, the rhetoric of dematerialization is already built-in in these device. Narratives of immateriality frame technology as a magic event, with the 'smartness' experienced as a seamless, even supernatural, occurrence. This can happen precisely because the materiality of the digital is hidden. With their emphasis on design and performativity, smart devices continue to hide the process of production, the exploitation and the power relations that have made them possible. But there is another aspect to consider: the compulsive tactility (all that swiping, tapping, pinching, scrolling) that smart objects demand from users produces a repetitive, obsessive, even morbid, fetishistic attachment to devices.

The fetishism of artificial intelligence (AI)

A case in point is AI. If we take AI as a particular kind of commodity, then here the process of fetishization reaches new heights.

First, AI is fetishized as a technology. Enveloped in dominant techno-deterministic and techno-positivistic discourses, AI becomes an autonomous force with its own instrumentality – and teleologically projected, like all other technologies. Second, AI relies on hidden digital infrastructures which maintain the illusion that this technology is truly 'artificial' – that is, unhinged from historicity, humanity and contingency. In this sense the fantasy of a self-generating AI with its propelling rhetoric of self-governing automation is the embodiment of textbook commodity fetishism. Not only are the producers and consumers of digital smart objects unrelated and indifferent to each other. More to the point, the fetishization of AI brings to its apex the dark future of capital envisioned by Marx.

As historian of science Simon Schaffer reminds us when he describes Charles Babbage's automatic machines, 'to make machines look intelligent it was necessary that the sources of their power, the labor force which surrounded and ran them, be rendered invisible' (Schaffer, 1994, p. 204). This original displacement of the labour behind machines is perpetuated by contemporary smart objects.

Commenting on the queues of customers camping overnight in front of Italy's soon-to-open largest Apple store (in Bologna), collective Wu Ming observes how 'putting the largest possible distance between upstream and downstream is the quintessential ideological operation under capitalism' (Wu Ming Foundation, 2011). It is this exploitative (and unseen) scaffolding that supports the soothing narratives of dematerialization, cloud and the internet as the ultimate phantasmagoria of unlimited communication, unbound opportunity and limitless freedom. Digital capitalism depends equally on this hidden labour and theoretical scaffolding as well as on the denial of this dependency.

Amazon Mechanical Turk, for instance, is a platform outsourcing to humans what computers cannot do very well (i.e. image labelling and classification). Its menial, repetitive labour shows that the intellectual, cognitive and immaterial – Post-Fordist – labour that creates the software needed to power our smart objects has virtually no value without the – still very Fordist – labour needed to manufacture hardware. 'Without factory workers and their labour, no valorisation of digital commodities, no Apple stock quote would be possible' (*ibid.*).

As the smooth surfaces of our laptops and our smartphones seduce us into an increasingly seamless, frictionless and increasingly infantile experience of interaction, they obfuscate the work and the resources that enter into its production. It is important to remember that these resources are both human and non-human: bodies, energies, materials, minerals, lives, time. This has an effect at once enrapturing and mystifying. As streams of data flow incessantly across our black screens, the world seems to be magically present before us, and made for us. What we no longer grasp in this spell-binding unmediated reality are the forces of production, the property relations, the debt and the profit, the relations of power, in one word, the profoundly violent asymmetry that moulds the digital experience appearing on screen, which is then cloaked by a truly mystical, supernatural, magic character.

An example of this is how the smart home is conventionally advertised through the rhetorical devices of magic and enchantment. In a commercial for the Beko connected Home[7] that would appeal to Harry Potter's fans, soothing music accompanies the images of two children playing in and with a smart environment that appears to be orchestrated by them but in reality catalogues and orchestrates the way they play, interact and, of course, consume. It is telling that the human delivery employee is made almost invisible. 'Smartified' domestic appliances showcase their functioning by in-real-time indexicality through metrics and icons that even a child can read. Smart objects must communicate with the users at all times. The fetishization of the home environment as 'smart' – that is, as responsive, pre-emptive, subservient – disguises under its reassuring and enchanting cloak a purposeful 24/7 tracking machine.

The aura of digital objects

Digital objects possess a specific kind of aura. They are predicated on the illusion that they are pieces of everyday magic. Their aura is a hiding mechanism that

---

[7] Retrieved 3 March 2021 from https://www.youtube.com/watch?v=cJmA6eXZmAg.

coalesces the rift between the product and everything/everybody that was necessary to its production (human, labour, earth resources, capital, exploitation). The 'superficially mysterious, perfect nature of the digitally manufactured, its magical aura—works to obscure the underlying physical reality of the digital and its subservience to human choices and agency' (Betancourt, 2015, p. 180).

In his critique of the inexorable march of automation, Michael Betancourt (*ibid.*, p. 34) talks about 'the bifurcation between design and facture, one where the devaluation of human labor reaches its apogee: rendered obsolete by the machine, there is no longer any need for human agency once the autonomous factory has been built except to switch it on'. Betancourt argues that in the automation of the modes of production of digital objects, human labour involved in production thins out, and only the human-as-designer remains as the only aspect of non-machine agency. For him, it is inherent in the ideology of automation that 'the productive human population appears obsolete, parasitic, on the "designers" whose plans they formerly executed' (*ibid.*, p. 34).

The totalizing discourse of automation, then, on the one hand tends to conceal the very material labour involved in its making – both in the lifecycle of digital objects (from the miners and the metal-scavengers, to the Silicon Valley janitors, Deliveroo and Uber platoons servicing digital elites) and in the endless labour of content generation. On the other hand, the endless search for anticipatory design and the ultimate seamless interface pinpoint to the increasingly subservient role of design within the techno-digital oligopoly – framed by techno-deterministic narratives about 'neutral' technology.

For Betancourt, digital capitalism should be called 'agnotologic capitalism' – a capitalism based on the systemic production and maintenance of ignorance, designed to confound, misconstruct and misinform. This agnotological order is fundamental to digital capitalism as it generates self-sustaining fictional value bubbles (e.g. dotcom, subprime property) while maintaining a narrow and strictly patrolled horizon for the social networks agents and their production of immaterial assets.

This is a process that constrains potential, pre-empts choices and predicts the future on the ground of past behaviours. It is what philosopher Antoinette Rouvroy aptly calls the regime of 'algorithmic governmentality' (2016, p. 6). It is this agnotological order that 'maintains its grip on the social: managing the emotional states of the consumers who also serve as the labor reserve is a necessary precondition to the management of the quality and range of information' (Betancourt, 2015, p. 207). Again, the extraction of time, attention, life – the biopolitical paradigm of extractive capitalism – is systemically enabled by the hidden structures that keep social agents in their simultaneous roles as users, consumers, producers and designers perpetually occupied, enduringly productive and caught in the digital snare 24/7. As we shall see in the next section, Marx already foresaw this metamorphosis of labour into both product and force of production well over 160 years ago.

Did Marx foresee automation?

There is a section in Marx's unfinished rough-drafted notebooks called *Grundrisse* (literally 'outlines') where the German thinker sketched some remarkably prescient ideas concerning the role of machines in the future development of capitalism (Marx, 1973).[8] What Marx lucidly describes in this section (known as the 'Fragment on Machines') is the way in which the machine engulfs and subsumes labour. The worker's skill set and even virtuosity honed in the course of many hours are no longer relevant. It is not the worker that makes the machine function; rather it is the machine that uses the worker's own labour as its raw combustible. 'It is the machine which possesses skill and strength in place of the worker, is itself the virtuoso, with a soul of its own' (Marx, 1973, p. 693).

Like workers need food to sustain themselves, so the machine's perpetual motion and animation are sustained through a constant ingestion of raw material. To Marx's list of *matières instrumentales* (coal, oil, etc.) we ought to add the coltan, niobium, tantalum and all the minerals needed to power digital devices and smart objects. But another raw material must be added to the list: the life power materialized in the continuous data stream that contemporary digital machines demand, and harvest from, their users.

If we substitute the word 'worker' with the word 'user' in the quotation below, we have accurate description of life in the universe of digital capitalism: 'The worker's activity, reduced to a mere abstraction of activity, is determined and regulated on all sides by the movement of the machinery, and not the opposite' (Marx, 1973, p. 693). As labour becomes 'a conscious organ, scattered among the individual living workers at numerous points of the mechanical system' (*ibid.*, p. 693), it is absorbed by the machine, which turns into a living system made of 'insignificant' individuals. Marx already saw how the transformation of the means of living labour into a mere living accessory of machinery signals a specific moment of the development of capital – the moment when labour becomes both product and the force of production. Marx anticipated that this metamorphosis of labour, far from being an accidental moment in the history of capital, is its foremost necessary tendency, which effectively reshapes labour into a new form adequate to its needs.

Seen through the lens of our contemporary smartified landscape, this is an accurate description of the triangulation user-producer-consumer we saw earlier, which turns time, attention and life itself into labour in order to extract value from them. The 'Fragment on Machines' is also where Marx discusses the notions of 'social brain' and 'general intellect'. As living labour becomes absorbed by the machines,

[8] Written in the winter 1857–8, these workbooks were lost for many years and were published for the first time in German in 1953, although a limited edition was published in 1939 in Moscow. Only in 1971 a partial version was made available in English.

capital sustains itself by the continuous 'accumulation of knowledge and of skill, of the general productive forces of the social brain' (Marx, 1973, p. 694). Not only do machines become the most appropriate form of capital; most significantly it is the general social knowledge (i.e. the social brain or general intellect) that becomes the force of production and product. This subsuming of the general intellect to the needs of the machine dictates how social life is constructed. How we live, the very conditions of human existence, in its multitude of social forms, are all dictated by what the machine demands.

The enforced multitasking that brings a user to be force-labour, product and producer comes with a remarkable paradox, lucidly explained by media theorist Geert Lovink (2019), that is, the seemingly pathological indifference of contemporary digital users to relinquishing data for the sake of speed, convenience and instant access. This is not a matter of ignorance (or not reading terms and conditions). It is rather a profound lack of concern that prompts (us) users to sign up for profiles knowing perfectly well what it is traded in the process. As the value of commodities resides (beyond their material sensible qualities) in their exchange value, for Marx the very soul of the commodity far from tangible becomes a ghost, a spectre. In this sense digital commodities are the culmination of the process of fetishization.

Conclusion

Marx notes how the progression of automation is ostensibly due to science and technology growing capacity at the service of capital enterprise. However, the pivotal element is the manoeuvre of 'dissection that through division of labor gradually transforms workers operations into more and more mechanical ones, so that at a certain point a mechanism can step in into their places' (Marx, 1973, p. 704). Mechanized, automated, atomized, workers' lives are appropriated by the machines in their entirety and in a 'coarsely sensuous form; capital absorbs labor into itself –"as though its body were by love possessed"' (*ibid.*).[9] In his famous 'Manifesto of Machinism' Italian designer Bruno Munari wrote,

> Today's world is a world of machines. We live among machines, they help us with everything we do in our work and recreation. But what do we know about their moods, their natures, their animal defects, if not through arid and pedantic technical knowledge?
>
> Machines reproduce themselves faster than mankind, almost as fast as the most prolific of insects; they already force us to busy ourselves with them, to spend a great deal of time taking care of them; they have spoiled us; we have to keep them clean, provide them with nourishment and rest, continually attend to them and meet their every need. In a few years' time we will become their little slaves.

[9] The quotation concluding Marx's paragraph is from Johann Wolfgang von Goethe's 'Faust'.

Munari wrote these words over eighty years ago and he was well aware of the danger of fetishizing technology. Indeed, he then goes on saying that machines must become works of art, and as a counterpoint to the Futurist adoration of machines he then started building his famous useless machines. Useless machines 'do not make anything, they do not eliminate labour, they do not save time and money, and they do not produce any commodities … [they are] objects to look at in the way one looks at a drifting group of clouds after spending seven hours inside a factory full of useful machines,' he says. Munari's useless machines intend to provoke us, poetically and imaginatively, to think about not only how we interact with smart objects but also how we can reimagine our whole relationship with them.

This paradox of automation was already evident to Marx. In spite of the reduction of human labour required by machines, there is never 'extra' disposable time for idleness, rest and personal growth. Instead, more and more (indeed all) time is plunged back into maximizing production. Workers (aka users) no longer own their own time. Machines have appropriated it. Users, in turn, have become slaves of the very machines they have created.

Bibliography

Astor, M. (2017, 25 July). Your Roomba may be mapping your home, collecting data that could be shared. *New York Times*. Retrieved 26 September 2019 from https://www.nytimes.com/2017/07/25/technology/roomba-irobot-data-privacy.html.

Avent, R. (2018, 27 June). A digital capitalism Marx might enjoy, *MIT Technology Review*. Retrieved 30 April 2020 from https://www.technologyreview.com/s/611480/a-digital-capitalism-marx-might-enjoy/.

Betancourt, M. (2015). *The Critique of digital capitalism: An analysis of the political economy of digital culture and technology*. New York: Punctum Books.

Crawford, K., & Joler, V. (2018). Anatomy of an AI system. Retrieved 26 September from https://anatomyof.ai/.

Fisher, E. (2015). How less alienation creates more exploitation? Audience labor on social network sites. In C. Fuchs & V. Mosco (Eds), *Marx in the age of digital capitalism* (pp. 180–203). Leiden: Brill.

Fuchs, C. (2014). *Digital labor and Karl Marx*. New York: Routledge.

Fuchs, C. (2017). Marx's capital in the information age. *Capital & Class*, *4*(1). 51–67.

Fuchs, C. (2018). Karl Marx & communication @ 200: Towards a marxian theory of communication. *TripleC: Communication, Capitalism & Critique*, *16*(2), 518–534.

Fuchs, C. (2019). Karl Marx in the age of big data capitalism. In D. Chandler & C. Fuchs (Eds), *Digital objects, digital subjects: Interdisciplinary perspectives on capitalism, labor and politics in the age of big data* (pp. 53–71). London: University of Westminster Press.

Fuchs, C., & Sandoval, M. (Eds). (2014). *Critique, social media and the information society*. New York: Routledge.

Fuchs, C., & Fisher, E. (Eds). (2015). *Reconsidering value and labor in the digital age*. Basingstoke: Palgrave Macmillan.

Fuchs, C., & Mosco, V. (Eds). (2015). *Marx in the age of digital capitalism*. Leiden: Brill.
Hornborg, A. (2012). *Global ecology and unequal exchange: Fetishism in a zero-sum world*. London: Routledge.
Hornborg, A. (2014). Technology as fetish: Marx, Latour, and the cultural foundations of capitalism. *Theory, Culture and Society*, *31*(4), 119–140.
Kreitmair, K. (2018). The seven principles for ethical consumer neurotechnologies. *The Neuroethics Blog*. Retrieved 26 September 2019 from http://www.theneuroethicsblog.com/2018/04/the-seven-principles-for-ethical.html.
Kreitmair, K., Cho, M., & Magnus, D. (2017). Consent and engagement, security, and authentic living using wearable and mobile health technology. *Nat Biotechnol*, *35*, 617–620. Retrieved 3 March 2021 from https://www.nature.com/articles/nbt.3887?platform=hootsuite#citeas.
Kroker, A. (2004). *The will to technology and the culture of nihilism: Heidegger, Nietzsche and Marx*. Toronto: University of Toronto Press.
Lovink, G. (2019). *Sad by design: On platform nihilism*. London: Pluto.
Marx, K. (1973). The 'Fragment on Machines'. In *Grundrisse: Foundations of the critique of political economy* (pp. 690–712). London: Penguin.
Marx, K. (1981). The fetishism of the commodity and its secret. In *The capital: A critique of political economy*, Volume 1, Part 1, Chapter 1, Section 4 (pp. 163–177). London: Penguin.
Marx, K., & Engels, F. (2012). *On religion*. Mineola, NY: Dover Publications.
Mason, P. (2018). Why Marx is more relevant than ever in the age of automation. *New Statesman*. Retrieved 26 September 2019 from https://www.newstatesman.com/culture/2018/05/why-marx-more-relevant-ever-age-automation.
McLellan, D. (2000). *Karl Marx selected writings*. Oxford: Oxford University Press.
McNally, D. (2011). *Monsters of the market: Zombies, vampires and global capitalism*. Leiden: Brill.
Mezzadra, S., & Neilson, B. (2017). On the multiple frontiers of extraction: Excavating contemporary capitalism. *Cultural Studies*, *31*(2–3), 185–204.
Munari, B. (1937). Manifesto del Macchinismo. *Arte Concreta*, *10*, 15 December, Milano. Retrieved 3 March 2021 from https://www.panarchy.org/munari/munari.html.
Negri, A. (2017). *Marx and Foucault*. Cambridge: Polity.
Pietz, W. (1993). Fetishism and materialism: The limits of theory in Marx. In W. Pietz & E. Apter (Eds), *Fetishism as cultural discourse* (pp. 119–151). New York: Cornell University Press.
Rose, X. (2017). Marxism 2.0: New commodities, new workers?. *International Socialism*, *154*. Retrieved 3 March 2021 from http://isj.org.uk/marxism-2-0-new-commodities-new-workers/.
Rouvroy, A. (2016). The digital regime of truth: From the algorithmic governmentality to a new rule of law. *La Deleuziana*, *3* (with Bernard Stiegler). Retrieved 3 March 2021 from http://www.ladeleuziana.org/2016/11/14/3-life-and-number/.
Sandoval, M. (2015). Foxconned labor as the dark side of the information age: Working conditions at Apple's contract manufacturers in China. In C. Fuchs & V. Mosco (Eds.), *Marx in the age of digital capitalism* (pp. 350–395). Leiden: Brill.
Schaffer, S. (1994). Babbage's intelligence: Calculating engines and the factory system. *Critical Inquiry*, *21*(1), 203–227.

Terranova, T. (2000). Free labor: Producing culture for the digital economy. *Social Text, 63*(18-2), 33–58.
Tucker, G. (2019, 4 August). Konnie Huq interview: Alexa has made family life with Charlie Brooker just like Black Mirror. *Sunday Times*. Retrieved 26 September 2019 from https://www.thetimes.co.uk/article/konnie-huq-interview-alexa-has-made-family-life-with-charlie-brooker-just-like-black-mirror-2nw7plcvp.
Wittel, A. (2012). Digital Marx: Toward a political economy of distributed media. *TripleC: Communication, Capitalism & Critique, 10*(2), 313–333.
Wu Ming Foundation (2011). Fetishism of digital commodities and hidden exploitation: The case of Amazon and Apple. Retrieved 3 March 2021 from htpp://www.wumingfoundation.com/giap/.
Wurman, P. R., Barbehenn, M. T., Verminski, M. D., Mountz, M. C., Polic, D., Hoffman, A. E., *et al.* (2016). System and method for transporting personnel within an active workspace. *US 9,280,157 B2* (Reno, NV, filed 4 September 2013 and issued 8 March 2016). Retrieved 3 March 2021 from http://pdfpiw.uspto.gov/.piw?Docid=09280157.

10 NOT A RESEARCH AGENDA FOR SMART OBJECTS

Ann Light

Smart objects and the Internet of Things (IoT) have not captured the consumer market, despite take-up in industrial contexts. Using feminist theory as the theoretical basis for a personal account of a multiply-failed research agenda, I describe twenty years of attempting to engage ordinary people in designing and planning for a smart, connected future. In the process, I consider the methodologies and the ethics that have contributed to my concern for the democratic aspects of networking everything.

Introduction

Sometimes things find unexpected uses. That happened for home security systems during the Californian 'Holy Fires' in 2018. During those days, the facility for remote access to IoT-enabled security through phones became the means of watching forest fires encroach on people's land. Owners were able to talk directly to the emergency services as fire fighters came to evacuate householders. They could reassure them there was no one home. This spared the services beating down doors and freed them for other duties. But it was shocking enough to make the international news.

Far away, I read about it in the paper (Wong, 2018). It was painful to learn of people watching flames inch closer to their houses, waiting. Images from the video systems made it viscerally real. Yet, as an HCI researcher, the repurposing of the security systems also intrigued me. The systems were conceived with human visitors in mind: used for remote package delivery and admitting cleaners. We can design tools, but we do not know how their use will play out. Watching flames approach on a phone was terrible for the owners, yet it gave them a chance to protect their property from the fire service's axes.

The other reason I was reading about these Californians is they are part of the (small) group of people who have installed a remotely accessible video home security system. Most people worldwide do not have such systems. When considering how smartness is changing homes and practices, it is also important to look at what is not, or should not be, happening (Baumer & Silberman, 2011). For most people, it doesn't matter if a thing is smart or artificially intelligent; it will find a place when it is affordable and useful. Only a few buy something for its technical novelty. As UK energy companies are learning in trying to roll out smart meters, sometimes you cannot introduce technical novelty even by giving it away (e.g. Meadows, 2017).

This chapter addresses smart objects by considering the reluctance of the consumer market to embrace smartness and my years of work on this lack of interest. I explore what did not work as expected in projects in which I had a personal stake, acknowledging the potential for investing ordinary objects with extraordinary powers and speculating on why this is not happening. In doing so, I stay alive to issues of social justice and care, acknowledging we live in an increasingly unstable world, irrespective of local ambitions to live smartly and safely.

Methodology

This chapter is written as a first-person account, even a memoir. It is a reflection on twenty years of practice and academic contribution. That said, speaking reflexively from experience is supported by a strong literature on situated knowledges (Haraway, 1988; Rose, 1997). My critical reflections owe commitment to this work, informed by a feminist understanding of what we can know and say (Light, 2011a, 2018; see Simandan, 2019, for a summary). And, in choosing to challenge the theme of this edited collection, I adhere to another commitment: to cross, queer and trouble (Light, 2011a).

Gillian Rose opens her discussion of positionality and reflexivity with a personal failure (1997). Not only does she give away the power of the objective author (after Haraway, 1988), but she deliberately undermines her authority as expert with a story that shows her failing to interpret information (1997, p. 306). She also shows, in her analysis, that this failure belongs to all of us. I hope to use failure here to question dominant narratives and agendas for smartness. Lack of success can be a personal or methodological failing as much as contextual inappropriateness. Only by understanding underlying factors can we generalize and learn. I present my experiences in that spirit, heeding calls to report what has not worked. But it is not a lonely failure. I have been phoned by commercial agencies to ask what the future of smart objects for the home may be, since they cannot find the market any more than I could find the application.

A lack of take-up

What follows, then, are several snapshots of this failed research agenda, problematizing aspects of bringing 'smartness' to ordinary life.

Early years

My engagement with smart technologies and networked products began in the late 1990s with concern at what I saw as unresponsiveness to the challenges of use. Mark Weiser's vision of a connected world of multiple screens was taking hold, promoted heavily by IBM (Kinsley, 2010). Weiser (1991) argued for computers 'invisible in fact as well as in metaphor'. Pointing to light switches, thermostats, stereos and ovens, he suggested these and more would be 'interconnected in a ubiquitous network' (1991, p. 98).[1]

I had finished a PhD on people's understanding of Web interactivity (Light & Wakeman, 2001) and I could see incongruities and vulnerabilities in this scenario. I identified a need to 'stop thinking about the product's end-users and start thinking about the system's end-designers … as we choose and use network components' (Light, 2002, in Rowland, Goodman, Charlier, Light & Lui, 2015, p. 626).

And so I became a would-be campaigner concerning a future I had been privileged to glimpse. Another ex-teacher and I formed a lobby group called *Transform-Ed* to consider how education could address the impact of connected things to give people more power of choice over what they enabled. Our fear was that people would only understand the potential of what became 'smartness' too late to share in its conception or protect themselves from its misuse. Whereas, we argued, people understand how to rearrange physical objects, the invisible data layer in data-enabled objects was problematic. Yet, just as TV remote controls had allowed one family member convenient means to override others' viewing, there would be new power relations.[2] It would undo 'the democracy of the visible switch' (Rowland *et al.*, 2015).

We realized that, at this time, those with a grasp of this potential had mostly had exposure to computer science. But, unlike advocates for computational thinking

[1] Ubiquitous, calm and pervasive computing eventually became smart objects and IoT.

[2] Smart home protocols allow dominant members to set lighting and heating regimes (Rowland *et al.*, 2015) and even spy on ex-partners (e.g. Small, 2019).

FIGURE 10.1 The old Transform-Ed comic graphic from 2004, designed to explain a not-yet-existent, invisible problem.

(Wing, 2009), we did not seek to teach coding. The crux of our concern was political, part of an ongoing interest in how digital networks (re)structure society (e.g. Light, 2011a; 2019). Ours was an early concern with what is now being made a sociotechnical agenda in some curricula (e.g. Dindler, Smith & Iversen, 2020).

But there were no data-enabled goods on hand at the point of *Transform-Ed* (2004–5). We were worried ahead of their production that a failure to understand networks, data flows and the potential for aggregation would disadvantage the public. While this was a fair concern, it was also a reason for indifference to the issue. Computer scientists were sceptical of our points about social engineering and others were unaware of what was possible. Another reason we had little impact, I acknowledge, is that our work was largely to rationalize, not publicize, our arguments. As a lobby, we were out of our depth. In the end, the main output of *Transform-Ed* was an image that I still use (Figure 10.1), because its message has never lost its relevance.

DemTech (2007–8)

The *Democratising Technology* project (DemTech, funded under the *Designing for the 21st Century* programme; Inns, 2009) was a direct legacy of these concerns. Written as a successful funding application in 2006, I intended to establish a method for engaging publics in the scrutiny of networks. Our premises included

> that digital networks herald significant and hard-to-grasp changes in technology; that new pockets of marginalization will result from this; that older people represent a good cross section of society and yet a statistically

> marginalized group; that everyone has something to offer the design process; that overcoming exclusion is as much about values as skills. (Light, Simpson, Weaver & Healey, 2009)

DemTech became known for launching a group of older men on a quest to design a water turbine. It is testimony to what arts methodology can do for bored older people (*ibid.*; Clarke, Briggs, Light & Wright, 2016) and a good example of design with communities (Simonsen & Robertson, 2012). But, it can also be seen as a series of failures to engage with networked technologies.

A research team came together around Lois Weaver's performance development techniques for creating dramatic *alter egos* (Weaver & Shaw, 2007), which we hoped to adapt to address networked designs. Tackling the theme of digital networks was new to Lois, but her methods worked well to reassure participants (East End Londoners in their later years) and engage them in speculation through imagining futures in character. What was less clear was how to introduce the networked angle.

The first failure was dramatic. In our first workshop, we reached a point when a mystified old lady stands holding one end of a series of ribbons while researchers variously hold other ends. She has been asked what the threads represent to her and, even with coaching, she remains baffled. Envisaging networks is not working.

We do better as we go along. The Geezers is a club for retired men, most having worked in skilled manual labour and, at the time, not internet users. When they imagined potential networked products as part of DemTech exercises, these included a teleporting device and a virtual holiday. But the striking feature was their desire to reuse old skills and knowledge to innovate in the field of renewable technology, not networks.

While participants had become conversant with the idea of networked objects, they had not become interested in them.

At the end of the project, we held an exhibition and a public symposium about participation in design. The Geezers showed their plans for a water turbine (Light *et al.*, 2009). Another participant (then in her seventies) spoke about her DemTech experiences, increased confidence and buying a laptop to support her group (Light, 2011b). The exhibition and symposium were well attended, but interest focused on inclusion for older people, not issues of incipient networking. What I christened *SINET* at outset, standing for 'The Social Implications of Networking Everything' (the term 'IoT' had not yet emerged), was still of no interest to the public. And the many write-ups and progressions of the project, still going strong after twelve years, also focus on other aspects.

In other words, the participants found their own meaning in our work. This came with positive effects: for instance, in October 2014, The Geezers launched a prototype water turbine in The Thames outside the British parliament (Clarke *et al.*, 2016). These older people were no longer excluded from design discussions, showing value to a series of techniques that treated participants as experts on life experience, social relations and the ethics of technology. Yet, as to embracing or critiquing networked futures, it is clear their hearts lay elsewhere. The DemTech team failed to make networked objects sexy, as targets of love or hate, and participant priorities remained family and friends, well-being (including environmental concerns), and fulfilment through mastery/use of skills, which, it turned out, did not require networked artefacts. So be it. It is not a failure I regret.

Contemporary times (2013–20)

In 2013, I became technical editor for a book on designing connected products (Rowland *et al.*, 2015). It was fascinating to be back in this terrain after a period researching social and ecological sustainability, but less had changed than I thought. There were still few meaningful applications. (The book took off only when a limited form of smart object industry was finally established.)

Nonetheless, the slow crawl towards application did not stop an industry's dreaming. Smart objects might not be popular, but smart infrastructure had fans.

I tracked these imaginaries. As well as subverting the concept of *smart city* – 'Smart for whom?' – with the *more-than-human* smart city (Heitlinger *et al.*, 2018b) – 'What of foxes and weeds?' – I looked at the sterility of the images that accompanied smart city materials. 'Search on "Smart City" with Google Images,' I told students. (*Perhaps you are trying that search now?*) 'What do you notice?'

Unless things have progressed, you will find images that are semi-schematic, many with data symbols marked over them. They are often night-time scenes and most resemble the Manhattan skyline. Trees and green spaces are almost absent and there are few people, fewer animals and only a couple with bike lanes. They look as if all information is aggregated and all control centralized. They are not resilient to power outages. They do not resemble a world in which I would like to own/use smart objects.

In contrast, some design challenges these visions, such as Heitlinger's connected seed bank (Heitlinger, Bryan-Kinns & Comber, 2018a) and watering can (Heitlinger, Bryan-Kinns & Jefferies, 2014), which share tales, recipes and growing advice at a communal farm. These smart objects resist the 'City' clichés; they are human-scale and mindful of other species. They build a sense of local belonging: not efficiency or exploitable data, but something rooted in life.

Discussion

In writing this chapter, I notice that, nearly twenty years on, I am still critiquing emerging narratives rather than full-blown take-up. Implementation of smart infrastructure, such as supply-chain management (making small producers more vulnerable and reducing workforces (Light, 2010)) has been rapid, but this reflects a prioritization of efficiency that mostly stops with delivery. As the industry puts it, 'the smart home market is stuck in the "chasm" of the technology adoption curve, … struggling to surpass the early-adopter phase and move to the mass-market phase' (Business Insider, 2016). Phones and televisions are "smart", but used as/for media. Smart speakers operate timers and play music rather than contributing to business as predicted (eMarketer, 2020; Smith, 2020). The gap remains between the potential of data-enabled 'smart' objects and what people make of them.

Perhaps there is more to consider than the usual adoption concerns such as newness, cost and so forth. The invisibility of data and its movements is not just an ethical issue; it is a reason that people cannot see the point of these tools. Infrastructure is, by definition, invisible (Star, 1999) and networks tend to be infrastructural. Do people feel *at home* with technology 'largely determined by its hidden, distributed and networked aspects' (Reddy, 2018, p. 171)? Even 'network' seems too concrete a description of the ether round smart objects. I am reminded of Zygmunt Bauman's *fluid forms of assembly* (2000) and Peter Sloterdijk's *foam* (2016).

During many years of research, I learnt that most people focus on things, not networks. Looking at *what is between* things is not intuitive. When attention was drawn to links, it quickly refocused. Similarly, data are treated as something solid, seen in terms of information, not instruction, so media and news-gathering remain primary functions, rather than process commands that link objects. And 'smartness' continues to be a mysterious term.

Unfortunately, there remain ethical issues that emanate from this obscurity. Beyond publicized security issues are more nebulous matters of privacy and control, which, being context dependent, are harder to elaborate. Data travel and combine; even local household dynamics are potentially affected by new data streams. As yet, there is still no pervasive sense of the increased inequalities, disturbing power relations or social stereotyping such tools can introduce. And there is no discussion of over-reliance on single providers, or the vulnerabilities of locking a whole house or city into a remote monolithic corporation for vital services. Indifference to politics accompanies indifference to function.

On reflection, engaging people in designing and planning for a smart, connected future has not worked well as either a political or practical endeavour. But, as Bauman says, "the job of restoring to view the lost link between objective affliction and subjective experience, has become more vital and indispensable than ever" (2000, p. 211). In the process of trying, I have learnt more about inclusive methodology, but this has not led me to a greater sense of how – or, more importantly, why – one might strive for connected futures. The conclusion is that there are more important research agendas. One agenda concerns exploring the dangerous techno-bubbles of people making things for the sake of it; another addresses tools for what people actually care about. And smart objects seem particularly ill-suited to matters of care (kith/kin, fulfilment, well-being), rather than matters of measurement and control. Although the chapter starts with a highly emotional example, even this use is an unusual application of monitoring and managing. Monitoring and managing are actions from a safety, not a care, paradigm. And smart objects are only able to provide for our safety if they can respond to changing circumstances, rather than lock us into sterile, even infantilizing, regimes.

Crisis has a way of changing things fast, and just as dances and coffee mornings have gone online during the Covid-19 pandemic, the case for handling things remotely is being made by circumstance, not research. The politics of design change as the opportunities do. Some of the world will come out of the pandemic with altered sensibilities about how to use technology and may find new purpose for the trade in data that networked goods require. Yet, if we are concerned with/for the democratic aspects of networking everything, then the mismatch between techno-commercial imaginaries and people's hopes and needs must be acknowledged. These tensions should shape the future of this research agenda or curtail it as an unnecessary extravagance in an uncertain world. If they teach us anything, crises, chaos and scarcity point towards more flexible, caring and forgiving systems, not just 'smarter' ones.

Conclusion

As a writer, editor, activist and researcher, my reflections on the challenge of considering connectivity, data flows, privacy and future living have highlighted what cannot be achieved through user research, even while dominant media and research tropes play into understandings of potential.

The nature of personal reflection is that it cannot cover a field in breadth or depth but instead offers an intimate picture, a slice determined by perspective. As a chapter sitting alongside others in this book, it acts as a commentary: A claim to question stories of progress and trouble achievements, drawn from twenty years of contrary examples. These are not my only examples of considering the Social-Impact-of-Connecting-Every-Thing. This account was written particularly to speak of my concern with ongoing technical agendas as populations do not embrace smartness in the way that technocrats would welcome and be able to capitalize upon.

The challenges continue. I am only a little disappointed as a researcher. I am simultaneously grateful for the resistance of domestic spaces (unlike public, commercial and political spaces) to the smother of data, the 'California Ideology' of breaking at speed and the proposal that connectivity surrounds but circumvents the person, as body, mind and collectivity. I am interested by the indifferent consumer, the sceptical activist and the relative uncontrollability of the citizen. I want to advocate for other, care-based and interdependence-promoting, values.

However, I close by thinking again of the homeowners watching fire through their home security systems. I am writing as a virus has changed social relations more profoundly than anything in my lifetime. In a world that is increasingly unstable and given to unpredictable change, uses for smartness are still evolving. The chance to control things remotely looks welcome, but we must never stop asking who has that control. And, if we are to stay healthy, connected and in balance with an evolving planet, how do data-enabled objects help?

Bibliography

Bauman, Z. (2000). *Liquid modernity*. Bristol: Polity Press.

Baumer, E. P. S., & Silberman, M. S. (2011). When the implication is not to design (technology). In *Proceedings of CHI '11* (pp. 2271–2274), New York: ACM.

Business Insider (2016, 7 October). Amazon wants to bring Alexa outside of the Echo. *Business Insider*. Retrieved 20 June 2020 from https://www.businessinsider.com/amazon-wants-to-bring-alexa-outside-of-the-echo-2016-10.

Clarke, R., Briggs, J., Light, A., & Wright, P. (2016). Situated encounters with socially engaged art in community-based design. In *Proceedings of DIS '16* (pp. 521–532). New York: ACM.

Dindler, C., Smith R. C., & Iversen, O. S. (2020). Computational empowerment: Participatory design in education. *CoDesign*, *16*(1), 66–80.

eMarketer (2020), Purchases Via Smart Speakers Are Not Taking Off, *eMarketer*, February 4th 2020. Available online: https://www.emarketer.com/content/purchases-via-smart-speakers-are-not-taking-off (accessed 20th June 2020)

Haraway, D. (1988). Situated knowledges: The science question in feminism and the privilege of partial perspective. *Feminist Studies*, *14*(3), 575–599.

Heitlinger, S., Bryan-Kinns, N., & Jefferies, J. (2014). The talking plants: An interactive system for grassroots urban food-growing communities. In *Proceedings of CHI'14* (pp. 459–462). New York: ACM.

Heitlinger, S., Bryan-Kinns, N., & Comber, R. (2018a). Connected seeds and sensors: Co-designing internet of things for sustainable smart cities with urban food-growing communities. In *Proceedings of PDC'18* (n.p.). New York: ACM.

Heitlinger, S., Foth, M., Clarke, R., DiSalvo, C., Light, A., & Forlano, L. (2018b). Avoiding ecocidal smart cities: Participatory design for more-than-human futures. In *Proceedings of PDC'18* (n.p.) New York: ACM.

Inns, T. (Ed.). (2009). *Designing for the 21st Century: Volume 2: Interdisciplinary Methods and Findings*. Aldershot: Gower.

Kinsley, S. (2010, 12 March). Ubiquitous computing: Mark Weiser's vision and legacy. *Spatial Machinations blog*. Retrieved 20 June 2020 from http://www.samkinsley.com/2010/03/12/ubiquitous-computing-mark-weisers-vision-and-legacy/.

Light, A. (2010). Bridging global divides with tracking and tracing technology. *IEEE Pervasive*, *9*(2), 28–36.

Light, A. (2011a). HCI as heterodoxy: Technologies of identity and the queering of interaction with computers. *Interacting with Computers*, *23*(5), 430–438.

Light A. (2011b). Democratising technology: Inspiring transformation with design, performance and props. In *Proceedings of CHI'11* (pp. 2239–2242). New York: ACM.

Light A. (2018). Writing PD: Accounting for socially-engaged research. In *Proceedings of PDC'18* (n.p.). New York: ACM.
Light, A. (2019). Redesigning design for culture change: Theory in the anthropocene. In P. Rodgers (Ed.), *Design research for change* (pp. 243–256). Lancaster: Lancaster University Press.
Light, A., Simpson, G., Weaver, L., & Healey, P. G. (2009). Geezers, turbines, fantasy personas: Making the everyday into the future. In *Proceedings of Creativity and Cognition '09* (pp. 39–48). New York: ACM.
Light, A., & Wakeman, I. (2001). Beyond the interface: Users' perceptions of interaction and audience on websites. *Interacting with Computers, 13*, 325–351.
Meadows, S. (2017, 2 August). Six reasons to say no to a smart meter. *Telegraph.* Retrieved 20 June 2020 from https://www.telegraph.co.uk/money/consumer-affairs/six-reasons-say-no-smart-meter/.
Reddy, A. (2018). Feeling at home with the internet of things. In L. Kronman & A. Zingerle (Eds), *The internet of other people's things: Dealing with the pathologies of a digital world* (pp. 171–181). Linz, Austria: Servus.
Rose, G. (1997). Situating knowledges: Positionality, reflexivities and other tactics. *Progress in Human Geography, 21*(3), 305–320.
Rowland, C., Goodman, E., Charlier, M., Light, A., & Lui, A. (2015). *Designing connected products: UX for the consumer internet of things.* Sebastopol, CA: O'Reilly Media.
Simandan, D. (2019). Revisiting positionality and the thesis of situated knowledge. *Dialogues in Human Geography, 9*(2), 129–149.
Simonsen, J., & Robertson, T. (Eds). (2012). *Routledge international handbook of participatory design.* London: Routledge.
Sloterdijk, P. (2016), *Foams: Spheres III.* (H. Wieland, Trans.). Cambridge, MA: MIT Press.
Small, T. (2019, 9 January). How smart home systems & tech have created a new form of abuse. *Refinery29.* Retrieved 20 June 2020 from https://www.refinery29.com/en-ca/2019/01/220847/domestic-abuse-violence-harassment-smart-home-monitoring.
Smith, C. (2020, 21 February). 17 amazing Amazon Alexa statistics and facts (2020). *DMR.* Retrieved 20 June 2020 from https://expandedramblings.com/index.php/amazon-alexa-statistics.
Star, S. L. (1999). The ethnography of infrastructure. *American Behavioral Scientist, 43*(3), 377–391.
Weaver, L., & Shaw, P. (2007). Make something: A manifesto for making performance. In E. Aston & S. Case (Eds), *Staging international feminisms* (pp. 174–183). London: Palgrave Macmillan.
Weiser, M. (1991). The computer for the 21st century. *Scientific American, 265*, 94–104.
Wing, J. (2009). Computational thinking. *Journal of Computing Sciences in Colleges, 24*(6), 6–7.
Wong, J. C. (2018, 17 August). Californians watch wildfires burn their houses via home security cameras. *The Guardian.* Retrieved 20 June 2020 from https://www.theguardian.com/us-news/2018/aug/17/california-wildfires-livestream-security-cameras.

11 TOWARDS WISE OBJECTS: THE VALUE OF KNOWING WHEN TO QUIT

Pim Haselager

Constructive ethics considers the ethical, legal and societal implications (ELSI) of artificial intelligence (AI) in order to elucidate what will become possible (and when), what is desirable or what should be avoided and what should be regulated, by whom, when and how. An early identification of the concerns of stakeholders may help to identify a technology's main risks. Liability concerns are crucial to a developing technology. Smart objects (SOs) will increasingly operate in (1) dynamic and unpredictable situations, (2) with a variety of agents with different cognitive and behavioural capacities, (3) as part of teams with unclear responsibility transitions, (4) with users that can experience an inaccurate sense of agency and that (5) intentionally or unintentionally, may not always work towards objectively desirable goals. Therefore, I will make a plea for the development of 'wise objects': SOs that know when to protect their users by switching themselves off.

Introduction: On the ethical, legal and societal implications of smart objects

Developments in AI and cognitive neuroscience (CNS) take place with an ever-increasing speed. Their enormous effects on daily life practices and society at large (from communication practices of individuals to ways of working and earning wages) necessitate an investigation of their ELSI. It is important that such ELSI analyses do not stand 'outside of' the science and technology but rather aim to be interactive. This implies that ELSI analyses should be the result of, and provide input for, communication with scientists and designers and potential stakeholders about what is possible, desirable or avoidable. This way, constructive ethics is part of the research and design process, instead of merely providing ethical codes before, or evaluations after, research and development (R&D). Ethicists working in the domains of SOs would therefore, I suggest, do well to engage in a continuous cycle of listening to scientists and designers, analyse presuppositions or implications of ongoing R&D, inform practitioners and stakeholders of potential consequences, and ask them about possibilities for change or improvement. From this perspective, ethics is not about telling others what should (not) be done, nor to instruct researchers to 'be good'. Instead, it can clarify or raise issues that require reflection and stimulate discussion, the results of which, sometimes, can be integrated into research. This will lead not only to ethically more prudent technology but also to commercially more successful products. After all, applications that, through their design, forestall or avoid concerns of clients and stakeholders will be more acceptable to users.

For this reason, it is important to ask several basic ELSI questions as early as possible in the research and design process. The first question is simply: What is possible? This question is a complicated one because often is not easy to separate the hope from the hype, and because time frames are regularly not clearly specified. Regarding the first, issue, companies like Gartner (https://www.gartner.com/en/research/methodologies/gartner-hype-cycle) produce regular analyses of developments in the field of AI and neuroscience, that indicate the estimated location of developing applications in a so-called 'Hype cycle'. They suggest for a variety of developing technologies whether they have been just triggered, suffer from inflated expectations, go through a period of disillusionment because of not fulfilling these expectations or finally reaching a level of useful productivity. For instance, in 2016 the Internet of Things (IoT) was located at the peak of the inflated expectations, because it was estimated to be still far removed from becoming genuinely practically applicable, while being much discussed in popular media (https://blogs.gartner.com/smarterwithgartner/files/2016/11/Hype-Cycle-for-the-Internet-of-Things-2016_Infographic-01.png). Although, of course, there is room for argument or disagreement regarding the exact location of a particular technology (or even the exact meaning and extension of a particular label), the fact that much advertised or discussed technologies may contain a large amount of overpromise complicates a proper reflection on the ethical implications of such technologies. Just like the products they are about, ELSI analyses could be hyped too, in the sense that they exaggerate prematurely the risks associated with a certain type of application.

Moreover, it is important to specify the temporal path that is being considered in an ELSI analysis. Many discussions about the hopes, concerns or risks of a technology fail to indicate a clear timeframe they consider in discussing this technology and its associated risk assessment. Without attempting to be overly precise, it generally is useful to distinguish what is currently possible (actually existing applications) from what will be possible in the near future (currently discussed in research papers, with the potential to develop applications in the next 5 years). These, in turn, have to be separated from an estimate of what would be possible in the long run (say the next twenty-five years or so), which has to be distinguished again from fantasizing about what 'ultimately' might become possible (science fiction).

A second set of questions concerns the assessments of stakeholders regarding what is desirable ('dreams'), what should be avoided ('nightmares') and how the technologies involved should be stimulated or restricted, via funding, regulations or laws. Gupta, Fisher and Frewer (2011) indicate that perceived risk is one the most frequently investigated socio-psychological determinants of public acceptance of a technology, much more than trust or perceived benefit. This implies that a proper risk analysis of a developing or to-be-developed technology will be crucial, right from the start. I will come back to this issue in the next section.

A third set of questions revolves around the identification and consultation of stakeholders in technology, which can be a complex and time-consuming task. Just to give a small list of candidates to be considered in general: end users, researchers, scientific experts or organizations (e.g. in relation to self-regulation via ethical codes), patient groups, caregivers, policymakers (governments), legal institutions, non-governmental organizations, companies developing the technology, companies using the technology, insurance companies, the general public and undoubtedly many other candidates. It is for this reason that ELSI analyses should start early in the research and design process. Even though this runs the risk of ethically assessing technology that is not there yet, societal debates about what to pursue or avoid, about what to regulate and why, take time too. Waiting for the technology to be out on the market before societal debates about ELSI take place runs the risk of being too late to effectively diminish the potentially negative effects of the technology. One example that comes to mind the loss of privacy on the internet, which at least in part can be seen as a consequence of not discussing early enough, for example, the effects of tracking cookies on websites. Although serious attempts are now being made to restore some of the privacy lost, the difficulties involved in this endeavour clearly illustrate the danger of engaging in ELSI analyses too late.

On the creative use of SOs by different agents in dynamic environments

As indicated above, perceived risks are an important element in the evaluation of a developing technology. An important factor in the risk assessment of SOs is that they will operate in a dynamic, unpredictable environment. As the autonomy and intelligence of SOs increase, these products will be capable of dealing with increasingly more complex and demanding tasks. Using AI 'in the wild' (Hutchins, 1995) implies that the SOs will operate in messy and quickly changing surroundings. In general, it is hard if not impossible to foresee exactly how such SOs will behave in dynamic environments. In the field of robotics, for example, prototypes get tested extensively in lab renditions of real environments. But so far, it is, at least to my knowledge and experience, rarely the case that such robots are tested in real-life circumstances under unrestricted or uncontrolled conditions for such prolonged periods of time that a complete, full-scale assessment is possible. Simply put, the world is just too big, complex and dynamic to exhaustively test relatively intelligent and autonomous systems.

Second, the label 'Human-Robot Interaction' (HRI) does not do justice to the ecologies of SOs, robots and various other types of intelligent agents that will interact in real-life settings. Just taking an elderly care institution as an example, studies usually focus on how the elderly person will interact with the robot, and even, in some cases, the caregiver as well. But in real life it is to be expected that robots will have to interact with a variety of other agents, for example, elderly, adults, patients, children, animals and other robots, with each having their own individual and behavioural (in)capacities, interactive styles and preferences. Here, too, tests will run short by far of the complexities to be expected in real-life robot interaction with a variety of other agents.

Third, speaking of a 'caregiver' may be an oversimplification, given the various parties involved in care contexts that each, to some extent, may be responsible for and/or determining parts of the robot's behaviours. In many situations care is not the responsibility of an individual, but of a team. For instance, in a caregiving context, the institution may be the owner of the SOs (say, a robot), local technicians may help to setup the robot for the specific environment it should operate in, professional caregivers may be instructors as well as collaborators of the robots, while the elderly are end users who instruct and experience the effects of the robot's actions. It will not be immediately clear, or even difficult after the fact, which agents carry which parts of the responsibility for the effects of an SO's action.

Fourth, it need not always be easily determinable to which extent users of smart technology will be aware of their actual use of it, not even for those users themselves. Increasingly, smart technology moves towards becoming symbiotic systems, becoming combined with or encompassing their users, rather than being used by them. In the contexts of brain-computer interfacing (Haselager, 2013; Krol, Haselager & Zander, 2019), as well as shared AI-user control (Vilaza, Haselager, Campos & Vuurpijl, 2014; Abbink *et al.*, 2018), situations may occur where users are unaware that they are acting or are incorrect in their experience of a sense of agency. That is, users may misinterpret their contributions to the performance of an action, either underestimating or exaggerating their contributions.

Finally, it may be incorrect to assume that under ordinary conditions, the user will always have beneficial intentions regarding the use of the SO. In some cases, the potential negative consequences of the use may be unintentional. A classic example is that of a cat sitting on a Roomba, attacking a pit bull (https://www.youtube.com/watch?v=vf9wHkkNGUU). The function of this SO, that is, to clean the floor, was made more difficult (by the cat's weight) and its actions were to some extent appropriated by the cat for different purposes, or at least leading to unplanned behavioural improvisations, such as attacking the dog. In addition to such accidental improvised use, the behaviour of bystanders or users could even go intentionally against the SO's functions, or the SO itself. This has been observed, for example, in the mal-education of chatbot 'Tay' (Vincent, 2016) that deliberately was taught to make racist and sexist statements, or in cases of robot bullying (Salvini *et al.*, 2010), where humans attack and damage robots. In an increasingly smart ecology, it seems rather likely that unforeseen, unplanned or even intentional 'creative use' with potentially negative consequences will manifest itself.

These five aspects of 'creative use' of SOs may have legal consequences, for example, in assessing the liability for the consequences of an SO's actions. For instance, Asaro (2012, p. 171) has pointed out the complexities of applying product liability in the context of smart systems, in his case, robotics:

> Legal liability due to negligence in product liability cases depends on either *failures to warn*, or *failures to take proper care* in assessing the potential risks a product poses. ... What constitutes proper care, and what risks might be foreseeable, or in principle unforeseeable, is a deep and vexing problem. This is due to the inherent complexity of anticipating potential future interactions, and the relative autonomy of a robotic product, once it is produced.

In tort law, strict liability concerns the imposition of liability on a party without a finding of fault (negligence or intention). The law imputes strict liability to situations it considers to be inherently dangerous: defective products, dangerous tools. Here the question arises how to evaluate SOs working within a dynamic and unpredictable environment (including e.g. cats, pit bulls and potentially agents with bad intentions as in the examples above). Should or could such an SO be considered as potentially 'defective' or 'dangerous', at least under some circumstances?

A different but related issue comes to the fore when one considers Article 6 of the EU Directive (85/374/EEC; see also Dodds-Smith, 2017), which states,

> A product is defective when it does not provide the safety which a person is entitled to expect, taking all circumstances into account, including:
>
> (a) the presentation of the product;

(b) the use to which it could reasonably be expected that the product would be put;

(c) the time when the product was put into circulation.

Just how much safety is a person entitled to expect regarding smart products, and what uses can SOs be reasonably expected to be put to? It does not seem that a framework for answering these questions in relation to the development of SOs exists. Yet it is unlikely that standard ways of addressing these questions within the contexts of traditional, non-smart objects will easily be translatable to this new domain. Perhaps even more important than directly answering these questions is the question whether the field of SOs is *aware* of such liability issues. To what extent are such considerations part of the research and design processes?

Midas's touch and cars with parachutes

The consequences of being unaware (or neglecting) issues surrounding responsibility for SOs and potential product liability can be vital. Awareness of these consequences is important not just for the development of safe products but also for the societal acceptance of these products, or even for the legitimate introduction of such products on the market. Hence, I suggest that objects should be more than 'smart'. What we need are 'wise objects' in the sense of carrying within their design a control mode related to responsibility and liability concerns. Smart technology may be said to be in danger of having the so-called Midas touch, in that it, once operative, may affect everything *continuously*. As Turkle (2008, p. 2) indicates, 'We are tethered to our "always-on/always-on-us" communication devices ... always ready-to-mind and hand.' This raises questions, once the technology is on, who can turn it off again, when and how? Should the system itself be equipped with some kind of 'emergency brake'? What would be the most graceful way of building emergency brakes into the system?

This presents a challenge that can be perceived as paradoxical, especially by engineers and designers. In the first stages of developing a product, all the attention and work is generally directed at creating and improving the SOs main function (whatever it may be). Yet, from an ELSI perspective, it is precisely during those early stages that one should think about its complete opposite, namely preventing it from continuing to do whatever it is doing by turning it off. Just as a simple example, when designing the motor of a car, it is logical to focus on improving the motor, because driving fast(er) is obviously the car's main function. Yet, from the perspective of stakeholders (ranging from drivers to passengers and other traffic participants to insurance companies and legislative institutions), the capacity to stop might be considered to be even more important than the capacity to drive. The paradox implies that as soon as one has thought of a rudimentary implementation of the basic function of the designed object, one needs to start addressing its opposite: a functionally easy to use, 100 per cent reliable, off-switch. Ignoring this issue might lead to suboptimal additions to the design later, for example, a car with only a parachute as its main brake.

In several cases, of course, risks of a product or technology are apparent enough to be taken into account from the start. Early on in the design process of cars, the need to be able to stop at any time was too obvious to be ignored, preventing later ad hoc patches like parachutes. However, it is far from clear that in the case of SOs operating in increasingly real-life environments, in an increasingly autonomous way, such risks are as apparent as in the case of cars. Hence, it is vital that serious attention (and some creativity) is applied to considering the types of risks SOs might bring, and the types of emergency brakes that are required to diminish such risks. Indeed, just as an illustration, within the EU, debates have started about the requirement for 'kill switches' on robots (Kottasova, 2017). Part of the challenge of designing wise objects lies in the development and acceptance of a framework for addressing risk, and brake-related issues. How can a technology be stopped as immediate and as fail-proof as possible? Who should be able to decide when to turn it off? Who would be held accountable for turning it off (or not)? Such questions are not to be seen as (later) additions to, but as an integral part of, the design process to be addressed and solved early on.

Wise objects

One option that presents itself in the context of SOs is that they are designed such that instead of (or preferably in addition to) having reliable ways to be stopped, SOs could be designed to turn off themselves. Just like a wise person knows when to shut up, a wise object should know when to stop. Simple examples exist of robots that can or should say 'No' to their users (Briggs & Scheutz, 2015; Förster, Saunders & Nehaniv, 2018; Peeters & Haselager, 2019). To what extent would it be possible to design SOs that can say 'No' to themselves? This might be important when human intervention would be too late to be meaningful, as the damage may already have been done, or too late to be effective, as when a rogue system spinning around at high speed would make an emergency brake unreachable (Arnold & Scheutz, 2018). It could be even more relevant to have self-terminating SOs in cases where their users require protection in relation to, for instance, consequences that the user is not, or insufficiently, aware of. This plea for wise objects is part and parcel of the broader perspective of beneficial AI (e.g. Russell, 2017).

Is it possible, for instance, to design SOs that collect data in order to provide better services to block data transfer for further processing (or even delete the data) when local analysis (within the SO) indicates a privacy risk? Ideally, wise objects should function as virtuous guardians of a user's privacy instead of the data-sucking vampires that they currently often are. Obviously, there are many challenges here. What criteria would have to be used, what threshold settings would function reasonably enough often enough? Which application domains require such wise objects most urgently? SOs for young children, for example, smart toys, might provide an interesting domain for further study.

Wise objects are SOs that can be trusted. As van den Brule *et al.* (2014, 2016) indicate, trust can be defined as the willingness of one agent (the 'trustor') to be vulnerable to the actions of another agent (the 'trustee'). The trustor depends on the trustee to reach its goals, but there is risk to the trustor if the trustee's actions fail or betray. In current SOs the risks of failure and/or betrayal can be unacceptably high. The most pressing challenges in design therefore don't involve the development of increasingly smarter objects, but rather of objects that are wise.

Conclusion

A consideration of the typical ELSI questions regarding what is possible when, and what possibilities should be strived for or avoided, quickly leads designers to concern themselves with the potential risks of their products. In the case of SOs significant risks derive from five factors inherent in their aimed for functionality. The more smart and autonomous objects will be, the more they will function for prolonged periods of time in dynamic and unpredictable circumstances, dealing with increasingly complex tasks. In those situations, it is likely that the SOs will have to interact with not just one user but with changing ecologies of users, SOs and various other types of agents with different cognitive and behavioural (in)capacities. Because the design, setting up, training, collaboration and use of SOs often involves many different agents, attributing (aspects of) legal responsibility may become quite complex. Sometimes, human users may not be fully aware of their (lack of) agency when using smart technology. In other cases, the use of SOs can be 'creative', in the sense of unintentionally or intentionally working towards negative consequences in not easily foreseeable or humanly preventable ways. I have therefore suggested that it is important to focus, from the start of the design process, on working towards wise objects: SOs that know when to quit.

Bibliography

Abbink, D. A., Carlson, T., Mulder, M., Winter, J., Aminravan, F., Gibo, T., *et al.* (2018). A topology of shared control systems: Finding common ground in diversity. *IEEE Transactions on Human-Machine Systems, 48*(5), 509–525. doi: 10.1109/THMS.2018.2791570.

Arnold, T., & Scheutz, M. (2018). The 'big red button' is too late: An alternative model for the ethical evaluation of AI systems. *Ethics and Information Technology, 20*, 59–69.

Asaro, P. M. (2012). A body to kick, but still no soul to damn: Legal perspectives on robotics. In P. Lin, K. Abney & G. A. Bekey (Eds), *Robot ethics: The ethical and social implications of robotics* (pp. 169–186). Cambridge, MA: MIT Press.

Briggs, G., & Scheutz, M. (2015). 'Sorry, I can't do that': Developing mechanisms to appropriately reject directives in human-robot interactions. In *Proceedings of the 2015 AAAI Fall Symposium on AI and HRI.*

Brule, R. van den, Dotsch, R., Bijlstra, G., Wigboldus, D. H. J., & Haselager, P. (2014). Do robot performance and behavioral style affect human trust? *International Journal of Social Robotics, 6*, 519–531.

Brule, R. van den, Bijlstra, G., Dotsch, R., Haselager, P., & Wigboldus, D. H. J. (2016). Warning signals for poor performance improve human-robot interaction. *Journal of Human-Robot Interaction, 5*(2), 69–89.

Dodds-Smith, I. (2017, 21 May). Recent developments in European product liability. *Arnold&Porter*. Retrieved 22 October 2020 from https://www.arnoldporter.com/en/perspectives/publications/2017/05/recent-developments-in-european-product-liability.

EU Council directive 25-07-1985, (85/374/EEC). Retrieved from https://eur-lex.europa.eu/legal-content/EN/ALL/?uri=CELEX%3A31985L0374.

Förster, F., Saunders, J., & Nehaniv, C. L. (2018). Robots that say 'no': Affective symbol grounding and the case of intent interpretations. *IEEE Transactions on Cognitive and Developmental Systems, 10*(3), 530–544.

Gupta, N., Fischer, A. R. H., & Frewer, L. J. (2011). Socio-psychological determinants of public acceptance of technologies: A review. *Public Understanding of Science, 21*(7), 782–795.

Haselager, W. F. G. (2013). Did I do that? Brain–computer interfacing and the sense of agency. *Minds & Machines, 23*(3), 405–418.

Haselager, P., Mecacci, G., & Wolkenstein, A. (forthcoming). Can BCIs enlighten the concept of agency? A plea for an experimental philosophy of neurotechnology. In O. Friedrich, A. Wolkenstein, C. Bublitz, R. J. Jox & E. Racine (Eds), *(Clinical) neuroethics meets artificial intelligence: Philosophical, ethical, legal and social implications*. Cham: Springer.

Hutchins, E. (1995). *Cognition in the wild*. Cambridge, MA: MIT Press.

Kottasova, I. (2017, 12 January). Europe calls for mandatory 'kill switches' on robots. *CNN*. Retrieved from https://money.cnn.com/2017/01/12/technology/robot-law-killer-switch-taxes/index.html.

Krol, L. R., Haselager, W. F. G., & Zander, T. O. (2020). Cognitive and affective probing: A tutorial and review of active learning for neuroadaptive technology. *Journal of Neural Engineering, 17*(1). https://doi.org/10.1088/1741-2552/ab5bb5.

Peeters, A., & Haselager, W. F. G. (2019). Designing virtuous sex robots. *International Journal of Social Robotics, 13*, 55–66.

Russell, S. (2017). Provably beneficial artificial intelligence. In *The next step: Exponential life* (pp. 178–192). BBVA OpenMind. Retrieved from https://www.bbvaopenmind.com/en/books/the-next-step-exponential-life/.

Salvini, P., Ciaravella, G., Yu, W., Ferri, G., Manzi, A., Mazzolai, B., *et al.* (2010). How safe are service robots in urban environments? Bullying a robot. In *19th International Symposium in Robot and Human Interactive Communication* (pp. 1–7). Viareggio, Italy: IEEE.

Turkle, S. (2008). Always-on/always-on-you: The tethered self. In James E. Katz (Ed.), *Handbook of mobile communication studies* (pp. 121–137). Cambridge, MA: MIT Press

Vilaza, G. N., Haselager, W. F. G., Campos, A. M. C., & Vuurpijl, L. (2014). Using games to investigate sense of agency and attribution of responsibility. In *Proceedings of the 8th*

Brazilian Games and Digital Entertainment Symposium (SBGames), Porto Alegre (2014, Nov 12–14) (pp. 393–399). https://www.sbgames.org/sbgames2014/papers/culture/full/Cult_Full_Using%20games%20to%20investigate.pdf.

Vincent, J. (2016, 24 May). Twitter taught Microsoft's AI chatbot to be a racist asshole in less than a day. *The Verge*. Retrieved 22 October 2020 from https://www.theverge.com/2016/3/24/11297050/tay-microsoft-chatbot-racist.

INDEX